Two Arguments for the Identity of Indiscernibles

Two Arguments for the Identity of Indiscernibles

GONZALO RODRIGUEZ-PEREYRA

OXFORD
UNIVERSITY PRESS

Great Clarendon Street, Oxford, OX2 6DP,
United Kingdom

Oxford University Press is a department of the University of Oxford.
It furthers the University's objective of excellence in research, scholarship,
and education by publishing worldwide. Oxford is a registered trade mark of
Oxford University Press in the UK and in certain other countries

© Gonzalo Rodriguez-Pereyra 2022

The moral rights of the author have been asserted

First Edition published in 2022

Impression: 1

All rights reserved. No part of this publication may be reproduced, stored in
a retrieval system, or transmitted, in any form or by any means, without the
prior permission in writing of Oxford University Press, or as expressly permitted
by law, by licence or under terms agreed with the appropriate reprographics
rights organization. Enquiries concerning reproduction outside the scope of the
above should be sent to the Rights Department, Oxford University Press, at the
address above

You must not circulate this work in any other form
and you must impose this same condition on any acquirer

Published in the United States of America by Oxford University Press
198 Madison Avenue, New York, NY 10016, United States of America

British Library Cataloguing in Publication Data

Data available

Library of Congress Control Number: 2022939630

ISBN 978–0–19–286686–8

DOI: 10.1093/oso/9780192866868.001.0001

Printed and bound by
CPI Group (UK) Ltd, Croydon, CR0 4YY

Links to third party websites are provided by Oxford in good faith and
for information only. Oxford disclaims any responsibility for the materials
contained in any third party website referenced in this work.

Contents

Preface   vii

    Introduction   1

1. Preliminaries   5
2. Trivializing Properties and NT-properties   31
3. Black's World   61
4. The Possibility of Black's World   75
5. Two Arguments for PII   101

References   127
Index   133

Preface

Although my views on the topic of this book have changed and developed considerably in the last 26 years, the seed of this book is my MPhil dissertation, 'A New Argument for the Principle of Identity of Indiscernibles', submitted to the University of Cambridge in 1995. Thus, I should like to thank the dissertation's supervisor, Alex Oliver. I am also grateful to Edward Craig, Jane Heal, and Hugh Mellor, with whom I also discussed the dissertation. More recently, previous versions of parts of this book were read and discussed in 2019–21 at Universidad Nacional del Litoral, Université Paris Diderot, Université de Genève, Taiwan National University, University of Rochester, Oriel College, Universidad Adolfo Ibañez, the University of St Andrews, the University of Oxford, and the Central European University. I thank the audiences in question and, particularly, Sophie Allen, Fatema Amijee, Maria Rosa Antognazza, Sophia Arbeiter, Richard Arthur, Bharam Assadian, Ralf Bader, Riccardo Baratella, Don Baxter, Elisa Besençon, Tomasz Bigaj, Philipp Blum, Samuel Boardman, Martina Botti, Tim Button, Gabriel Catrén, Matthew Collier, Earl Conee, Fabrice Correia, Zachary Crouch, Gabriel Dumet, Sam Elgin, Vera Flocke, Juan Luis Gastaldi, Giacomo Giannini, Anil Gomes, Louise Hanson, Vera Hoffmann-Kolss, Katherine Hong, Nick Jones, Alexander Kaiserman, William Kilborn, David Mark Kovacs, Bridger Landle, Karol Lenart, Olimpia Lombardi, Ofra Magidor, Nikhil Mahant, Giovanni Merlo, Robert Michels, Paolo Natali, Matteo Nizzardo, Andrea Popescu, David Rabouin, Vincenzo De Risi, Bernhard Salow, Antonio Salgado Borge, Paolo Santorio, Jarred Snodgrass, Daniel Stoljar, Radivoj Stupar, Jan Swiderski, Antoine Taillard, Zach Thornton, Javier Vidal, Ed Wierenga, Nathan Wildman, and Christian Wüthrich. Various people read and commented on whole versions of the manuscript at different stages of its completion, and especial thanks go to them: José Tomás Alvarado, Paul Audi, Sebastián Briceño, Sam Cowling, Michael Della Rocca, Michael Townsen Hicks, Tien-Chun Lo, Martin

Pickup, Alex Roberts, Carlo Rossi, Maria Scarpati, Jonathan Schaffer, Al Wilson, and Ezequiel Zerbudis; and so also did Claudio Calosi, Bradley Rettler, and Erica Shumener as readers for the press. Claudio deserves a very especial mention since he read more than one whole version of the manuscript, and he is the person with whom I have discussed the ideas of this book the most. Last but not least, my thanks to Peter Momtchiloff for his interest in, and support of, this project from the beginning.

G R-P
October 2021

Introduction

The goal of this book is to present and advance two arguments for the Principle of Identity of Indiscernibles, the principle that says that no two objects can differ only numerically. The two arguments appeal to very different philosophical ideas: one is based on a version of the Humean principle that rules out necessary connections between numerically distinct objects and the other one is based on ideas about what grounds the having of certain properties by objects.

But this book also presents and advances two arguments against the Principle of Identity of Indiscernibles. This is because the Principle of Identity of Indiscernibles comes in many versions, and some are true while others are false. This is exactly what I shall argue here. In particular, there are two versions of the principle that I shall argue against: these are the version that rules out the possibility of intrinsically purely qualitatively indiscernible objects and the version that rules out the possibility of objects that are both intrinsically and extrinsically purely qualitatively indiscernible. There are already conspicuous arguments against these versions of the principle (Black 1952 and Adams 1979). But there is reason to present new arguments against these versions of the principle, since I think Adams' argument does not work (see Rodriguez-Pereyra 2017a) and the argument presented in Black's paper is subject to a powerful objection originally presented by Ian Hacking (1975). My arguments against those versions of the principle are persuasive, sound, and valid, and they are not subject to Hacking's objection.

Now, philosophers nearly always assume that, in order to be non-trivial, any version of the Principle of Identity of Indiscernibles must quantify only over purely qualitative properties or, as I prefer to call them, *pure* properties. *Impure properties*, often called non-qualitative properties and sometimes called impurely qualitative properties, are those such that having them consists in being related to one or more objects in particular; pure properties are those that are not impure. One of the main contentions of this book is that there is a non-trivial version of the Principle of Identity of Indiscernibles that quantifies over impure properties. And this is the version of the Principle of Identity of Indiscernibles for which I shall advance two arguments, based on the Humean and grounding considerations I mentioned above. I take the main contribution of the book to show that this is a non-trivial version of the Principle of Identity of Indiscernibles that has been neglected and for which there are compelling arguments.

Here is a brief synopsis of the book. The point of the first chapter is to lay out the elements I need to deploy my arguments for and against different versions of the Principle of Identity of Indiscernibles. Thus, in this chapter I shall introduce and explain several formulations of the Principle of Identity of Indiscernibles, and I shall explain my position on a range of issues, like objects and properties. I shall distinguish, for instance, between concrete and abstract objects, and between relational and non-relational, intrinsic and extrinsic, and pure and impure properties.

One important claim I make and argue for in Chapter 1 (Section 1.1) is that the principle that, necessarily, no two objects share all their properties, which is often considered a trivial version of the Principle of Identity of Indiscernibles, is indeed trivial but not really a version of the Principle of Identity of Indiscernibles. But this claim depends on a conception of the properties that trivialize the principle that, necessarily, no two objects share all their properties. Characterizing the class of trivializing properties, and distinguishing them from the non-trivializing properties, a complex and rarely attempted task, is the main object of Chapter 2. And once I have done that, it becomes clear that there is no trivial version of the Principle of Identity of Indiscernibles. *A fortiori*, the version of the principle I shall argue for is not a trivial version, although it is the weakest version of the Principle of Identity of Indiscernibles. Another important claim that will be argued for in Chapter 1 is that, contrary to what is always naturally assumed, the Principle of Identity of Indiscernibles is not the converse of the Indiscernibility of Identicals.

Now, it is often assumed that the point of the Principle of Identity of Indiscernibles is to ground numerical identity in the purely qualitative, but since the version of the principle I shall argue for quantifies over impure properties, it is compatible with the identity of objects being primitive—indeed, since I shall argue against the version of the principle that rules out the possibility of objects sharing all their purely qualitative properties, it follows that the identity of objects is not grounded in the purely qualitative. But, as I shall argue in Section 2.7, the Principle of Identity of Indiscernibles is not a principle about what grounds what, nor does it make any priority claims. It is rather a supervenience thesis according to which there can be no numerical difference without some extra-numerical difference. Nevertheless, I shall argue in Section 2.8, there is one sense in which the Principle of Identity of Indiscernibles can provide a principle of individuation.

After that, I am ready to proceed to my arguments. In Chapter 3 I shall argue that the spheres of Black's world are both intrinsically and extrinsically purely qualitatively indiscernible. This is necessary because some philosophers have argued that Black's spheres are not indiscernible in the relevant sense. And then in Chapter 4 I shall argue that Black's world is possible, which means that it is possible that there are objects that are both intrinsically and extrinsically purely qualitatively indiscernible. After rejecting three arguments for the possibility of intrinsically purely qualitative objects, I shall give my own argument that it is possible that there are intrinsically purely qualitative indiscernible objects (in the terminology to be introduced in due time, this is the claim that it is possible that there are objects sharing all their intrinsic pure properties). On the basis of that result, I shall then give an argument that establishes the possibility of Black's world, and therefore it establishes that it is possible that there are objects that are both intrinsically and extrinsically purely qualitatively indiscernible (in the terminology to be introduced in due time, this is the claim that it is possible that there are objects sharing all their pure properties, including both intrinsic and extrinsic pure properties). The arguments are new, and one important and interesting feature is that they do not appeal to an imaginative conception of the situation obtaining in Black's world, and therefore they are not vulnerable to the objection first made by Hacking in 1975. As I shall argue in Section 4.5, even if some of the premises or presuppositions of the second argument are rejected, it can be reformulated in such a way that it establishes, if not the possibility of Black's world, the possibility of objects sharing all their pure properties—and that is really my target.

Finally, in Chapter 5 I shall argue for the version of the Principle of Identity of Indiscernibles that rules out objects sharing all their non-trivializing properties, including impure ones. These are the argument from Humean considerations and the argument from grounding considerations I mentioned at the beginning. The arguments are largely independent from each other. But there is an interesting connection between them, which I shall point out in Section 5.5. Indeed, a consequence of the argument from grounding is that the version of the Humean principle I use in the other argument must be true. Nevertheless, since having two independent arguments makes my case more compelling, I provide independent reasons for the version of the Humean principle, reasons having nothing to do with the grounding argument. Finally, it is important to point out that although I shall largely focus my discussion on *concrete* objects, in the last section of the book, Section 5.7, I argue that the argument from grounding can be extended to abstract objects, and therefore that there are no objects, whether abstract or concrete, that share all their non-trivializing properties. Therefore, since, necessarily, every object is either abstract or concrete, the Principle of Identity of Indiscernibles is true of all objects whatsoever.

Although I find everything concerning the Principle of Identity of Indiscernibles fascinating and intriguing, there is a lot about it that I shall not discuss here—for instance, the status of the Identity of Indiscernibles in contemporary Physics, the relationship between the Bundle Theory and the Identity of Indiscernibles, Leibniz's arguments for the Identity of Indiscernibles and the use he makes of it in his philosophy, whether there can be purely qualitatively indiscernible universals, and a detailed discussion of Adams' argument against the Identity of Indiscernibles (except on the status of the Identity of Indiscernibles in contemporary Physics, I have written on the other topics in Rodriguez-Pereyra 2004, 2014, 2017a, 2017b, 2018). This is simply because this book is a single piece of argumentation towards supporting the weakest version of the Principle of Identity of Indiscernibles, and discussion of those and other topics is not necessary for my aims here.

1
Preliminaries

1.1 The Principle of Identity of Indiscernibles (PII hereafter) has a non-modalized version and a necessitated version. I shall focus on the necessitated version, simply because it is the logically stronger of the two. If two things or objects differ purely numerically, or merely numerically, they differ *solo numero*. So, here is my first statement of PII:

PII: Necessarily, no two objects differ *solo numero*.[1]

That there cannot be purely numerical difference means that numerical difference must always be accompanied by a difference other than numerical, which I shall call *extra-numerical difference*. Objects x and y differ numerically if and only if x and y are not one and the same object, or if and only if they differ with respect to *which* objects they are, while they differ extra-numerically if and only if they (also) differ in any way other than differing with respect to which objects they are. Since whenever objects differ extra-numerically they differ with respect to at least some property, PII is almost always, especially in the contemporary literature, formulated in terms of properties. Here is a common formulation:

(1) Necessarily, no two objects share all their properties.

For two objects to *share* a property is for both of them to have it; and for two objects to *differ* with respect to a property is for one of them to have it and the other one to lack it. So what principle (1) says is that necessarily,

---

[1] Indeed, Leibniz often formulates PII in terms of *solo numero* difference (see, for instance, Leibniz 2020: 14 and Leibniz 1967: 45; for a discussion of Leibniz's different formulations of PII see Rodriguez-Pereyra 2014: 15–25).

no two objects have exactly the same properties. Now, as is well-known, principle (1) is trivially true. For among the properties that objects have are what I shall call *properties of identity*, sometimes called *haecceities*, that is, properties like *being identical with Aristotle* and *being identical with Napoleon*. It is, of course, trivially true that no two objects can possibly share such properties. Thus, in order to discuss a substantive and interesting principle, philosophers exclude from the formulation of PII those properties that they think would trivialize it.

Although the triviality of (1) is noted in virtually every paper on the subject, what is rarely discussed is *why* certain properties trivialize it. My answer to this is that properties of identity, and other trivializing properties, are such that differing with respect to them does not require differing extra-numerically. I shall develop this answer in Chapter 2. But what is pertinent to note here is that since trivializing properties are those such that differing with respect to them does not require differing extra-numerically, principle (1) is not really a version of PII. Indeed, PII states that objects cannot differ only numerically, but principle (1) is consistent with objects differing only numerically, since Principle (1) allows objects to differ only with respect to those properties differing with respect to which does not require differing extra-numerically. Therefore, principle (1) is not a version of PII.

1.2 That principle (1) is not a version of PII does not mean that PII cannot be formulated in terms of properties. As I said, whenever objects differ extra-numerically they differ with respect to at least one property. In particular, whenever objects differ extra-numerically they differ with respect to at least one property such that differing with respect to it requires differing extra-numerically. Thus, this is my second formulation of PII:

> PII: Necessarily, no two objects share all those properties such that differing with respect to them requires differing extra-numerically.

Thus PII, unlike (1), contains a restriction in its quantification over properties, a restriction that excludes properties of identity and other trivializing properties. Since, as will become clear in Chapter 2, the properties such that differing with respect to them requires differing extra-numerically are the properties that do not trivialize (1), this is my third formulation of PII:

> PII: Necessarily, no two objects share all their non-trivializing properties.

This last formulation makes clear that PII is not a trivial principle. What is trivial is principle (1), but it should now be clear that principle (1) is not PII.

For the sake of abbreviation, I shall from now on call non-trivializing properties *NT-properties*. If so, we can regiment the formulation of PII in a mixture of English and the language of second-order logic thus:

> PII: Necessarily, $(x)(y)((F_{NT})(\text{if } Fx \text{ if and only if } Fy, \text{ then } x = y))$.

The four formulations of PII I have proposed are alternative but equivalent formulations of one and the same idea, and I shall explain the last formulation in more detail. One interesting aspect of the last formulation is that it makes clear that PII is not the converse of the Indiscernibility of Identicals—indeed, the converse of the Indiscernibility of Identicals is principle (1). That PII is not the converse of the Indiscernibility of Identicals is only a terminological misfortune, a terminological misfortune that should not mislead people into thinking that principle (1) should be identified with PII. For PII, whatever its name, is meant to be the principle that there is no difference *solo numero*. Such a principle, as we have seen, cannot quantify over all properties since it cannot quantify over properties of identity and other trivializing properties. But the Indiscernibility of Identicals quantifies over all properties. Therefore, PII is not the converse of the Indiscernibility of Identicals (that PII is the converse of the Indiscernibility of Identicals is assumed by everyone, mostly implicitly, but sometimes explicitly (e.g. Ladyman and Bigaj 2010: 117)). One consequence of this is that PII cannot secure, together with the Indiscernibility of Identicals, a definition of '$x = y$' as '$(F)(Fx$ if and only if $Fy)$'. But those who seek such a definition need not worry: given the Indiscernibility of Identicals such a definition can be secured by principle (1).

The modal adverb at the front of all four formulations of PII expresses *metaphysical* necessity. I cannot define or explain this notion here, and I shall assume a relatively good grasp of it on the part of the reader. But I should say that although I shall assume that what is logically necessary holds true in every metaphysical possibility, I shall make no assumption about whether the converse is true, that is, I shall leave open whether what is metaphysically necessary holds true in every logical possibility. Effectively, what I shall leave open is whether metaphysical necessity is absolute (for discussion of absolute necessity see Hale 1996 and 2015: 112–15). Thus, according to PII there is no metaphysical possibility in which two objects have exactly the same NT-properties.

The symbol '=' stands for numerical identity, and so '$x = y$' means that x and y are numerically identical, that is, one and the same object (more strictly, it means that, relative to each variable assignment, the object assigned as a value of 'x' and the object assigned as value of 'y' are numerically identical). The expressions 'if..., then' and 'if and only if' stand for the respective well-known propositional truth-functions.

The last formulation of PII has two first-order universal quantifiers and one second-order universal quantifier. The first-order quantifiers, (x) and (y), are intended to range over concrete objects—thus, the statement of PII incorporates an implicit restriction in the domain of the first-order quantifiers. By concrete objects I understand entities that are in space, time, or spacetime. They may be material or immaterial. Thus, the traditional examples of substances are typically concrete objects. But not every concrete object is a substance, not at least in fairly typical conceptions of substance. For instance, concrete proper parts of concrete objects are concrete objects too, but on many conceptions of substance, substances are simple, and on other conceptions of substance, the proper parts of substances are not substances. I take PII and its different versions to be discussed in this book as principles about concrete objects, in the broad sense of objects existing in space, time, or spacetime.

There is, of course, an interesting issue whether there can be indiscernible abstract objects, since although for some philosophers abstract objects cannot be indiscernible, for some others they can be.[2] In this

---

[2] For discussion of PII in relation to abstract objects see, among others, Ladyman 2005, MacBride 2006, Leitgeb and Ladyman 2008, De Clercq 2012, and Duguid 2016.

book I shall focus almost completely on *concrete* objects, and abstract objects will be mentioned or discussed only as it is necessary to advance or clarify a point about concrete objects, with the exception that at the end of the book, in Section 5.7, I shall argue that one of my arguments for PII also establishes that no abstract objects can be indiscernible.

One final point about my use of the word 'object': it might be possible for someone to take it to be definitional of objects that they satisfy PII. Although anyone can choose what to mean by one's words, I see little rationale in restricting the word 'object' to entities satisfying PII (Assadian 2019: 2558 makes a similar point). So, to be clear on this point, in this book I shall use the word 'object', unless preceded by the word 'abstract', to refer to *concrete* objects, including those concrete objects, if there are any, that fail to satisfy whatever version of PII I am discussing. And, to repeat, by concrete objects I mean those that exist in space, time, or spacetime. By abstract objects, I mean those that do not exist in space, time, or spacetime. What is my definition, then, of the general notion of object? I don't have a definition of this basic ontological category, although I do have an intuitive grasp of it, as most of us do. Fortunately, for my purposes in this work, I do not need to provide a definition, and so I shall leave it as an undefined notion. But although undefined, I can give the extension of my general notion of object: everything. Thus, its extension is the same as that of the notions of thing and entity, as they are commonly understood (and I shall occasionally use 'thing' and 'entity' in this maximally general way). Understanding the general notion of object in this maximally general way is not totally idiosyncratic (for a precedent see van Inwagen 2002: 180, 196). But although absolutely everything, whether abstract or concrete, particular or universal, a property or not, is an object, let me remind you again that in what follows my use of the unqualified word 'object' will refer to concrete objects.

What does the second-order quantifier, (F_{NT}), in the formulation above range over? I intend it to range over NT-properties—the point of the subscript is precisely to signal the restriction to this particular kind of properties. When a property is an NT-property is something to be discussed in detail in Chapter 2. But what are properties? Often they are taken to be universals, or tropes, or classes (sets). But here I shall take them to be, instead, predicable conditions, that is, conditions that could be predicated of an object. Here the word 'condition' is simply functioning as a noun applying only to whatever is predicable of an object (from which it follows that there is no need to worry about whether there are 'impredicable' conditions). In other words, properties are what meaningful predicates of objects predicate, or what could be predicated by a meaningful predicate of an object—the second disjunct is needed since there are more predicable conditions than there are meaningful predicates.[3] On this conception of properties, property F is the same as property G if and only if F and G are the same predicable condition. Since there are necessarily coextensive but different predicable conditions, properties are individuated hyperintensionally on this conception of them. And, of course, on such a conception of properties there are both contingently uninstantiated properties and necessarily uninstantiated properties. Several philosophers have maintained theories of properties like this (e.g. van Inwagen 2004: 131–8, Jubien 2009: 54–7, and Hale 2015: 37–40, to mention only three).

Now, to say that properties are what meaningful predicates predicate is not the same as saying that properties are the meanings of predicates. Saying the latter assumes that there are no other components in the meaning of a predicate than the condition it predicates. For instance, 'is a beast' (in one of its many meanings) predicates the same as 'is a non-human terrestrial mammal'—but do these two predicates have the same meaning? Not if the connotation of irrationality of 'is a beast' is part of its meaning. Whether predicable conditions exhaust the meanings of predicates is a difficult question that I am not in a position to discuss. Nevertheless, even if predicable conditions do not exhaust the meanings of predicates, it is still the case that to every meaningful predicate corresponds a predicable condition, the predicable condition it predicates. And this is what I take properties to be.

---

[3] A more natural way of expressing the second disjunct would be 'what a possible meaningful predicate stands for'. But although more natural, this is misleading, since it suggests the rather unnatural thesis that there are merely possible predicates, that is, possible but not actual predicates. Some welcome merely possible objects (I did so in my 2002, for instance, in the context of adopting Modal Realism as part of a defence of Resemblance Nominalism), and therefore they should offer no resistance to merely possible predicates. But in this work I shall not commit myself to merely possible objects. See Section 1.6 for more on this.

Note that on this conception of properties there are properties that are sometimes rejected, such as negative, disjunctive, conjunctive, and modal properties among others. This is because there are meaningful predicates that predicate negative, disjunctive, conjunctive, and modal conditions. Thus I accept such properties as *not being green, being green or being square, being green and being hot, possibly being square,* and *necessarily being the teacher of Aristotle.*

This is an abundant conception of properties, as opposed to a sparse one (the distinction is due to Lewis 1986: 59–60; see also Lewis 1983b: 345–6). On a sparse conception of properties, not every meaningful predicate corresponds to a property, since on that conception of properties they are what makes for qualitative similarity, they are intrinsic, highly specific, and there are as many as are needed to characterize things completely and without redundancy (Lewis 1986: 59–60; see Schaffer 2004 for an argument that Lewis' characterization in fact encompasses two different conceptions of sparse properties).

Lewis takes abundant properties to be sets (Lewis 1986: 60). Do I take them to be sets too? No. As I have said, I take them to be predicable conditions—and it would be a category mistake to think that a set is a predicable condition. A set cannot be predicated—although *being a set* can of course be predicated, but such a condition is not a set.

There are many reasons why it would be inadequate to identify abundant properties with sets in a work like this, but here is a particularly powerful one. The property of *being identical with Aristotle* is a trivializing property, but the property of *being the greatest philosopher ever* is an NT-property. Now, if properties are sets of their instances and Aristotle is indeed the greatest philosopher ever (a very plausible supposition, by the way), then the properties of *being identical with Aristotle* and *being the greatest philosopher ever* should be one and the same—but they cannot be, since one is a trivializing property and the other one is not. It is irrelevant to point out that although Aristotle may be the greatest philosopher ever, this is not necessarily the case. For even if it were necessary that Aristotle is the greatest philosopher ever, one of the properties would still be a trivializing property and the other would still not be, since, as we shall see in the next chapter, whether a property is trivializing has nothing to do with whether it necessarily applies to one and only one object.

As I said, I take properties to be predicable conditions. Are such conditions universals? Some of them seem to be, like the properties of *being green* and *being round*, since they can be multiply instantiated, or multiply satisfied, or multiply had. Others seem not to be, like the properties of *being identical with Aristotle* and *being the tallest man*, since they cannot be multiply instantiated, satisfied, or had. However, when I refer to universals in this work I always refer to universal concrete objects that are supposed to play the roles of accounting for the resemblance and causal powers of objects—universals thus understood are not predicable conditions. Immanent universals conceived in the manner of Armstrong are an example of universal concrete objects. But transcendent universals, even if they are not concrete objects, need not be seen as predicable conditions, and can also be taken to be what plays the roles of accounting for resemblance and the causal powers of objects (Alvarado 2020 develops such a theory). Similarly, I shall understand tropes as concrete objects that account for the resemblance and causal powers of objects—tropes thus understood are not predicable conditions either. Since I take properties to be predicable conditions, I do not take universals and tropes to be properties—at least not in the sense that I understand properties in this work.

1.3 The conception of properties I have adopted is abundant because I take every meaningful predicate to correspond to a property. Why have I chosen to use an abundant theory of properties rather than a sparse one?

As we have seen, PII quantifies only over NT-properties. Thus, one of my tasks in this book is to discuss and determine which properties are trivializing and which ones are not. To do this I have to discuss trivializing properties, but such properties, properties such as *being identical with Aristotle*, for instance, count as properties only on an abundant conception. Furthermore, a large class of NT-properties, properties such as *being a teacher of Aristotle*, also count as properties only on an abundant conception. Thus, choosing a sparse conception of properties would imply ignoring respects in which objects can differ extra-numerically.

These are reasons to choose an abundant conception of properties, but why have I chosen, in particular, a conception of properties on which they are predicable conditions? As we have seen, according to PII, any two objects must differ extra-numerically. But any extra-numerical difference will do: PII is not sensitive to the kind of extra-numerical difference that must accompany numerical difference. Thus, for the purposes of discussing PII, every predicate that expresses a condition with respect to which objects can differ should express a property, and predicates expressing different such conditions should express different properties. This is why I have adopted a conception of properties as predicable conditions.

It is sometimes interesting to discuss restricted versions of PII, versions that quantify only over sparse properties. But doing this does not force one to abandon the conception of properties as predicable conditions. For although sparse properties are typically identified with universals, or tropes, or resemblance classes, predicable conditions can also play the role of sparse properties. Indeed, some predicable conditions are highly specific, intrinsic conditions that mark qualitative similarity and such that there are as many of them as are needed to characterize things completely and without redundancy. Thus, by adequately further restricting the second-order quantifier of PII, one can formulate a version of PII according to which sharing all sparse properties entails numerical identity, and one can do this without abandoning the view of properties as predicable conditions.

1.4 Predicable conditions are abstract objects, since they are not spatiotemporally located. *Being red*, for instance, is neither in space, nor time, nor spacetime, though red objects of course are. Thus, the property of *being red* is an abstract object, not a concrete one. This generalizes to all properties, and therefore all properties are abstract objects, not concrete objects. In my last formulation of PII above I took the first-order quantifier to range over *concrete* objects. And all the versions of PII I shall discuss in this book will be principles about *concrete* objects. Thus, although one may raise the interesting question of whether any version of PII is valid for properties, this is not something I shall discuss in this book, except that, as I said, at the end of the book I shall argue that one of my arguments for PII applies to abstract objects and therefore to properties too.

Now, properties can be referred to by singular terms. For instance, I have already referred to a couple of properties by means of the singular terms 'the property of *being red*' and 'the property of *being the greatest philosopher ever*. This anti-Fregean stance has as a consequence that properties can be in the domain of first-order quantifiers.[4] And this might suggest that it is not necessary to formulate PII using a second-order quantifier. For it might be thought possible to appeal to the relation of *having* or *instantiating*, which obtains between any object and any property if and only if the property is truly predicable of the object, and use only first-order quantifiers in the formulation of PII, as follows:

PII: Necessarily, $(x)(y)(z_{NT})$(if x has z if and only if y has z, then $x = y$)

Although the third first-order quantifier ranges over properties in this formulation, this still formulates a principle about concrete objects since the first two first-order quantifiers in it are restricted to concrete objects: it asserts of *concrete* objects, and only of *concrete* objects, that in their case the sharing of a certain totality of properties entails numerical identity. Since I take properties to be abstract objects, this last formulation of PII is, as the previous ones, silent about whether the sharing of a totality of properties is sufficient for numerical identity in the case of properties.

Clearly, one *must* formulate PII in terms of three first-order quantifiers if one takes properties to be universals, tropes, or sets—although if one takes them to be sets, one should appeal to the relation of set-membership rather than the relation of *having* or the relation of *instantiating*. For universals and tropes are no more predicable conditions than sets, and second-order quantification is quantification into predicate position.

But is it correct to formulate PII using only first-order quantifiers, even if properties can be the values of first-order variables? It is true that

---

[4] There are those who reject that properties can be in the domain of first-order quantifiers (e.g. Jones 2016, Trueman 2021). I cannot go into this discussion in detail, but suffice to say that the view that properties can be in the domain of first-order quantifiers is also well supported; indeed, Bob Hale and Crispin Wright have views that imply that properties can be in the domain of first-order quantifiers, since for them properties can be referred to by singular terms (Wright 2001: 87 and Hale 2015: 31). But although for Wright this means that properties are objects, Hale avoids this consequence since for him singular terms need not refer to objects. Thus, I agree with both Hale and Wright in making properties first-order quantifiable, and I agree with Wright that properties are objects, but nothing I have said above commits me to their idea that the metaphysical categories of object and property ought to be explained in terms of syntactic categories (Wright 2001: 90 and Hale 2015: 32).

not every predicable condition can be the value of a first-order variable: 'does not instantiate itself' is a meaningful predicate, but what it predicates cannot be the value of a first-order variable, on pain of contradiction. Thus, it might be thought that objects that count as indiscernible when one quantifies over values of first-order variables need not count as indiscernible when one quantifies over values of second-order variables, since there is an extra property when one quantifies over values of second-order variables. If this is the case, the formulation of PII in terms of three first-order quantifiers is not equivalent to the formulation involving a second-order quantifier. But even if there is an extra property when one quantifies over values of second-order variables, all objects (that is, concrete objects) are alike with respect to the property expressed by the predicate 'does not instantiate itself', since no concrete object instantiates itself. Therefore, the predicate 'does not instantiate itself' does not provide a reason to think that objects that count as indiscernible when one quantifies over values of first-order variables need not count as indiscernible when one quantifies over values of second-order variables.

Nevertheless, I have two reasons to formulate PII in terms of a second-order quantifier. First, such a formulation makes the connection between properties and predicability more visible. Second, I am not in a position to rule out the possibility that there are objects that do not differ with respect to properties that are values of first-order variables but differ with respect to properties that are not values of first-order variables. Thus, I shall here adopt the formulation of PII using a second-order quantifier.

I have said a few times that since properties are abstract objects, PII is silent about whether the sharing of properties is sufficient for numerical identity in the case of properties. Now, universals and tropes, if they exist, are concrete objects, since they are supposed to exist in space and time, at least on some conceptions of them. Thus, PII applies to them and it entails that, if concrete universals and tropes exist, indiscernibility entails numerical identity in their case. But I shall have very little to say about concrete universals and tropes in this book. This is for two reasons. First, there is no reason to believe that concrete universals and tropes exist unless they are the entities playing the role of sparse properties. But, as I have argued elsewhere, concrete universals and tropes are not needed to play the role of sparse properties (Rodriguez-Pereyra 2002). Second, even if concrete universals and tropes exist, or even if they could exist, they pose no threat to my conclusion in this book. Indeed, at the end of the book the reader will be in a position to see that my arguments in Chapter 5 apply to them too.

1.5 I have already referred to properties by means of expressions of the form 'the property of F', where 'F' is the nominalization of a predicate, for instance the property of *being identical with Aristotle*, the property of *being red*, etc. But since properties correspond to predicates and predicates can be regimented as open sentences, there is another way of referring to properties, and this is by means of the property abstraction lambda-operator. The lambda-operator binds a variable from a first-order open sentence to designate the property expressed by that open sentence. For instance, given the open sentence 'x is red', the lambda-operator binds the variable to give '$(\lambda x)(x$ is red)', which designates the property of *being red*. Similarly, given the open sentence 'x = Aristotle', the lambda-operator binds the variable to give '$(\lambda x)(x$ = Aristotle)', which designates the property of *being identical with Aristotle*.

Note that, since properties correspond to monadic predicates, the open sentences corresponding to predicates expressing properties are open sentences with only one free variable. In subsequent chapters I shall often refer to properties by means of lambda-operator expressions.

1.6 I said that for two objects to differ with respect to a property is for one of them to have it and the other one to lack it. But since objects have and lack properties at times, it might be thought that what I need to say is that for two objects to differ with respect to a property *at a time t* is for one of them to have it at t and the other one to lack it at t. If so, since what PII states is that necessarily any two objects must differ with respect to at least one NT-property, PII would need a quantifier over times, and say that necessarily, for every time t, any two objects must differ with respect to at least one NT-property at time t. Such a version of PII would tolerate numerically different objects that are indiscernible *across times*.

But thanks to the abundant conception of properties I have adopted, this complication is unnecessary. For on the abundant conception of properties, objects have temporal NT-properties like *being red in 2017*, *visiting the Eiffel Tower in 2039*, etc. Note that the abundant conception of properties not only affords B-series NT-properties like the two properties I just mentioned, but also A-series NT-properties like *having been red* and *visiting the Eiffel Tower next year*.

This does not mean that I am committed to the respective irreducibility of the A-series and the B-series. I shall remain neutral in this work about which one reduces to which one, if either of them reduces to the other, or which one of them is fundamental, if only one of them is fundamental. My point is simply the following: whenever two objects differ because, say, one is red at time t and the other one is yellow at that time, but the former is yellow at another time t^* and the latter is red at t^*, they will differ across times with respect to some temporal NT-properties, whether it is properties like *being red in 2017* or properties like *having been red*, or properties of both kinds.

Thus, although objects must have and lack properties at a time, temporal properties allow cross-temporal differences to be registered in terms of differences with respect to temporal properties. There is therefore no need to incorporate a quantifier over times into PII. PII, as I have formulated it above, ranges over all NT-properties and therefore it ranges over temporal NT-properties, and so it rules out cross-temporal indiscernibility between numerically different objects.

But I must emphasize that this does not commit me to there being objects, numerically different or not, that differ across times. All I am committed to is that, if there are such objects, their differences can be expressed as a difference with respect to their temporal properties. I am not committed to there being objects that differ across times simply because I am not committed to there being objects that exist at different times. Indeed, my stance in this work is also compatible with presentism, although I am not committed to it either.

Similarly, the abundant conception of properties affords modal properties like *necessarily being human*, *possibly being a lawyer*, and *contingently visiting the Eiffel Tower*, and objects have, lack, and therefore differ with respect to, such properties. Thus, modal differences between objects can be expressed in terms of differences with respect to modal properties. But I shall not accept the existence of merely possible entities. Thus, I am here committed to Actualism, in the sense that what actually exists is all that exists. If so, although there could have been objects that differed from the ones that actually exist, there are no objects that differ across possible worlds (Actualism, as I have just defined it, leaves open the possibility that some actual abstract objects deserve the title of 'possible worlds'; I take no stance on this in this book, but even if that is indeed the case, there will be no (concrete) objects differing across possible worlds, since (concrete) objects are not parts of abstract possible worlds). That there are no objects differing across possible worlds does not mean, of course, that there are no modal differences between objects: the Eiffel Tower, for instance, necessarily coexists with the Eiffel Tower, while the Statue of Liberty only contingently coexists with it. But these modal differences can be captured in terms of differences with respect to modal properties: the Eiffel Tower has the property of *necessarily coexisting with the Eiffel Tower*, while the Statue of Liberty lacks it.

Indeed, I think Actualism is necessarily true, which means that it is impossible that there are merely possible entities. Although I believe that Actualism is necessarily true, an argument for or even a defence of Actualism is not possible in this work.[5] My commitment to Actualism, however, will have no bearing on the main arguments of this work. Despite my commitment to Actualism and my neutrality with respect to actual abstract possible worlds, I shall often put my points in terms of possible worlds, but this will only be for ease of expression and to make my points more effectively.

1.7 PII then says that if the non-trivializing predicable conditions that are true of a certain object x are exactly the same as those that are true of an object y, x and y are one and the same object. Now, properties are not the only predicable conditions: relations are too. For instance, one can predicate of two objects a and b that they love each other, or that they

---

[5] For the record, that I now find Actualism true does not mean that I have abandoned Resemblance Nominalism. For it now seems to me that Resemblance Nominalism can be developed without adopting Modal Realism. But that is a story for another occasion.

are two miles away from each other, or that the former is bigger than the latter. And objects can agree or differ with respect to their relations: *a* and *c* can be alike in that both of them are two miles away from *b*, but differ in that only *a* loves and is loved by *b*. So, since PII makes a connection between what is truly predicable of objects and their numerical identity, relational differences should not be ignored by PII.

One can bring relational differences to PII by taking it to be the principle that necessarily, no two objects share all their NT-properties, bear exactly the same NT-relations to the same objects, and are borne exactly the same NT-relations by the same objects. But since there are relations of n-adicity for any number n, a formulation of PII in terms of quantifiers and variables that allows for relations requires some extra complications.

Nevertheless, there is a more convenient expedient, which is to appeal to relational properties and let the second-order quantifier range over both non-relational and relational properties. Relational properties are those such that to have them *is* to be related to any object or objects—or, as I shall say, having them *consists in* being related to any object or objects (from now on when I say that something consists in something, what I mean is that to be the former is to be the latter).[6] For instance, if *a* and *b* are two miles away from each other, *a* thereby has the relational property of *being two miles away from b*; and if *a* and *b* are in love with each other, *a* thereby has the property of *loving b* and also the property of *being loved by b*. Similarly with relations of higher adicity: if *a* is between *b* and *c*, *a* thereby has the relational property of *being between b and c*.

It might be thought that there are relational differences that cannot be captured by differences in properties and therefore the expedient of using relational properties rather than relations in the formulation of PII compromises its adequacy. For in that case, the falsity of PII would not guarantee that objects need not differ extra-numerically, since it might be that, necessarily, every two objects must differ relationally in a way that cannot be captured by a difference in properties.

---

[6] My decision to use 'consists in' in this way has been influenced by conversations with Paul Audi.

Indeed, Quine (1960, 1976) famously distinguished three types of discernibility: absolute discernibility, which obtains between two objects if and only if only one of them satisfies an open sentence with one free variable; relative discernibility, which obtains between two objects if and only if there is an open sentence with two free variables that they satisfy only in one order; and weak discernibility, which obtains between two objects if and only if there is an open sentence with two variables that is satisfied by both of them in both orders but not by one of the objects with itself—in clearer words, weak discernibility obtains if and only if two objects bear a relation R to each other but at least one of them does not bear R to itself.

Philosophers often say or suggest that weak discernibility obtains if and only if two objects are related by a symmetric but irreflexive relation, or at least by an irreflexive relation (see, for instance, Saunders 2003: 293, Leitgeb and Ladyman 2008: 388–9, Ladyman and Bigaj 2010: 122, De Clercq 2012: 662, fn. 2, Muller 2015: 207). But neither the symmetry nor the irreflexivity of the relation is required for weak discernibility to obtain. We may say that weakly discernible objects are symmetrically related by a relation R and at least one of them is not reflexively related by R, provided we bear in mind that objects can be symmetrically related by non-symmetric relations and they can fail to be reflexively related by non-irreflexive relations. For the record, when Quine formulated weak discernibility ('discriminability', in the terminology of his 1976 paper), he did not require the symmetry or the irreflexivity of the relation in question. Indeed, he improved on Ivan Fox's criterion by removing the requirement that the relation in question be irreflexive: 'We can trim Fox's criterion a little by requiring not that his open sentence be irreflexive, but just that it be reflexively false of one of the objects to be discriminated.... A sentence in two variables *weakly discriminates* two objects if satisfied by the two but not by one of them with itself' (Quine 1976: 115–16).

Clearly, relative discernibility and weak discernibility apply only to objects that differ relationally, but it is important to note that even the first type of discernibility, absolute discernibility, can obtain between objects that differ relationally and even only relationally—for instance, two objects that differ only because one but not the other satisfies this open sentence with one free variable: 'Rxb'.

But using relational properties instead of relations in the formulation of PII does not compromise its adequacy. For no two objects can differ with respect to a relation without differing with respect to a relational property. This is because, for example, if two objects differ with respect to a dyadic relation there will be at least one formula of the form R*xb* or R*bx* that only one of them will satisfy, and from such formulas we can abstract the relational properties of *bearing R to b* and *being borne R by b*, at least one of which will be had by only one of the objects in question. Obviously, the case generalizes to relations of any adicity. Thus, letting relational properties fall in the range of the second-order quantifier of PII does not compromise its adequacy.

1.8 The distinction between non-relational and relational properties leads to another distinction between properties that will be very important in what follows. This is the distinction between purely qualitative and impurely qualitative properties or, for short, *pure* and *impure properties*. Here is my way of drawing the distinction: impure properties are those such that having them consists in being related in a certain way to a certain object or objects in particular, whether such object or objects are abstract or concrete; pure properties, on the contrary, are those that are not impure: having them does not consist in being related to any object or objects in particular—this might be because having them does not consist in being related in any way to any object or objects, or because although having them consists in being related in some way to something or another, having them does not consist in being related to any object or objects in particular.[7]

Sometimes the distinction between pure and impure properties is made in linguistic terms. For instance, Robert Adams proposes as a possible definition of purely qualitative properties that they are those

---

[7] My way of drawing the distinction between pure and impure properties is very similar to that of Khamara's, but there is a difference: for me the object or objects to which having an impure property consists in being related to, need not be concrete, and could very well be properties, while for Khamara such objects can only be spatiotemporal entities or space and time themselves and their parts (Khamara 1988: 143, 145). Sam Cowling (2015) has argued that all extant ways of accounting for the distinction between pure and impure properties are unsatisfactory and he proposes, on that basis, to take the distinction as primitive. Although his criticisms of the many ways of accounting for the distinction he discusses are correct, the characterization of the distinction I just gave is adequate, and so I do not share his primitivism about the distinction between pure and impure properties. See Plate 2022 for further discussion.

that could be expressed in a sufficiently rich language without the aid of referential devices like proper names, proper adjectives and verbs (such as 'Leibnizian' and 'pegasizes'), indexical expressions, and referential uses of definite descriptions (Adams 1979: 7—by the way, this is not Adams' preferred definition). Impure properties would be, according to this, those that cannot be expressed, in any language, without such referential devices. But there are at least two considerations that indicate that this is not the right definition of pure and impure properties. First, the reason why impure properties cannot be expressed without the use of such referential devices is that what partially determines whether an object has an impure property is *which* object or objects it is related to in a certain way, while pure properties can be expressed without such devices because which object or objects something is related to does not even partially determine whether it has a pure property (cf. Cowling 2015: 287). Second, impure properties can be expressed without the linguistic devices mentioned by Adams. For instance, they can be expressed by means of free variables under an assignment. If Aristotle has been assigned to the variable '*x*' then 'is a friend of *x*' expresses the property of *being a friend of Aristotle*, which is an impure property since it consists in being related in a certain way to a certain object in particular. That impure properties can be expressed by means of variables will be important in Section 5.7.

Let me give some examples to clarify the distinction between pure and impure properties. The properties of *being identical with Napoleon*, *being a teacher of Aristotle*, and *orbiting the Earth* are impure properties, since having them consists in being related in certain ways to certain objects in particular, namely Napoleon, Aristotle, and the Earth. But the properties of *being identical with something*, *being a teacher*, and *orbiting a planet* are pure, for although in order to have them an object must be related in certain ways to some objects, it need not be related in those ways to any objects in particular.

Similarly, the property of *being a teacher of the greatest philosopher ever* is a pure property too. One might point out that having this property requires being related in a certain way to the greatest philosopher ever, not just any philosopher whatsoever. True, but what is required to have this property is to be related in a certain way to whatever satisfies the condition by which the *relatum* is specified, that is, to be a teacher of whoever happens to be the greatest philosopher ever. This person happens to be Aristotle (or so I shall plausibly assume), but that it is Aristotle is not a requirement of the condition *being a teacher of the greatest philosopher ever*. Thus, the property of *being a teacher of the greatest philosopher ever* is a pure property. And note that this is not because it is possible, in whatever sense of possibility, for Aristotle not to have been the greatest philosopher ever—even if it had been necessary, in whatever sense of necessity, that Aristotle should be the greatest philosopher ever, it would still have been the case that being a teacher of Aristotle is not a requirement of the condition *being a teacher of the greatest philosopher ever*.

The distinction between pure and impure properties might be drawn in terms of dependence. The idea would be that having an impure property depends on the identity of one or more *relata*, while having a pure property does not depend on the identity of any *relata*.[8] In so far as one takes such dependence to be a necessary connection between having the property in question and being related in a certain way to the *relatum* or *relata* in question, such a characterization of the distinction between pure and impure properties is incorrect. For suppose it was the individual essence of Aristotle to be the greatest philosopher ever, and therefore it was metaphysically necessary and sufficient for being Aristotle to be the greatest philosopher ever. In that case whether or not one has the property of *being a teacher of the greatest philosopher ever* would depend on the identity of one's student—if one teaches Aristotle one has the property, if one does not teach Aristotle, one does not have the property. But the property of *being a teacher of the greatest philosopher ever* is a pure property!

Nevertheless, one can appeal to the notion of dependence to draw the distinction between pure and impure properties, provided one emphasizes that whether the dependence in question obtains or not has to do with requirements arising from the property in question itself. Thus, having the property of *being a teacher of Aristotle* depends on the identity of the person one teaches, while having the property of *being a teacher of*

---

[8] Cf. the view that draws the distinction in terms of grounding, discussed in Cowling 2015: 290.

the greatest philosopher ever does not depend on the identity of the person one teaches. For the property of *being a teacher of Aristotle* requires that whoever has it must teach Aristotle, but the property of *being a teacher of the greatest philosopher ever* does not require that whoever has it must teach Aristotle or any other person in particular—even if it is metaphysically necessary and sufficient for being Aristotle to be the greatest philosopher ever. Thus, one could distinguish between pure and impure properties by saying that impure properties are those such that having them depends on the identity of one or more *relata* because they require that whoever has them is related in a certain way to some object or objects in particular, while pure properties are those such that having them does not depend on the identity of any *relata* because they do not require that whoever has them is related in a certain way to any object or objects in particular.

Thus, when I say that impure properties are those such that having them consists in being related in a certain way to some object or objects in particular, what I mean is that it is a requirement of those properties that having them consists in being related in a certain way to some object or objects in particular. Correspondingly, it is not a requirement of pure properties that having them consists in being related in a certain way to any object or objects in particular.

All non-relational properties are also pure properties, obviously, since they do not consist in being related in any way to anything. Thus, the properties of *being red* and *being hot*, for instance, are pure, since they are non-relational. Here someone might object. Suppose one believes in universals (tropes) and thinks that what makes objects red is that they are related, through instantiation (having), to the universal redness (a certain particular red trope); or suppose one is a resemblance nominalist and thinks that what makes objects red is that they resemble *those* objects (namely the red ones). Some such people might object that a property like *being red* is in fact an impure relational property. But this is wrong. Properties are predicable conditions, and the condition of *being red* makes no reference at all to any relation to any universals, tropes, or other entities. It might still be the case that what makes objects red is being related in certain ways to certain entities (universals, tropes, red objects), but one must distinguish between what makes objects red, or makes objects have the property of *being red*, and what the property of *being red* is. That what makes objects red or makes them have the property of *being red* is a relational affair does not entail that the property of *being red* is a relational property—and it is not: the property of *being red* is the condition predicated by the predicate 'is red', and nothing there requires a relation to anything. Thus, properties like *being red* and *being hot* are non-relational and they are therefore pure properties.

The same point applies to properties that might be thought to consist in relations to kinds, for instance the property of *being water*. It might be thought that having these properties is being related in some way to the kind *water*. But this is not so, even if there is such a kind as *water*, and even if it is metaphysically necessary that everything having the property of *being water* is related in a particular way to the kind *water*. For given the condition predicated by the predicate 'is water', satisfaction of this condition requires simply to be watery, not that there is a kind, over and above watery things, to which watery things are related in a certain way.

One might think that, at least on the Kripkean understanding of natural kind terms, a property like *being water* is an impure property, since having it consists in being related to certain things, namely the members of the sample by means of which 'water' gets a reference. But this is not so. Although for Kripke a natural kind term gets 'defined' as 'the substance instantiated by the items over there, or at any rate, by almost all of them' (Kripke 1981: 135), he is clear that the items in the sample only fix the reference of 'water', but they do not form part of its meaning—indeed, there might still have been water even if none of those items had existed. So, even on the Kripkean understanding, what is predicated by 'is water' does not require for anything to satisfy it that it has to bear a relation to anything else in particular. Thus, even the Kripkean can view *being water* and similar properties as pure properties.

Some people believe that biological species are objects, and that being a member of a certain species is to be a part of the scattered object which is that species (for discussion see Ereshefsky 2017). If that is the case, the properties of *being a homo sapiens* and *being a canis familiaris* are impure properties, but they are also relational ones, since to have those properties consists in being related to certain objects. Of course, this does not mean that the properties corresponding to the everyday predicates 'being human' and 'being a dog' are impure properties—in my view such properties are non-relational conditions and therefore they are pure properties.

A similar point applies to the properties of *being red*, *being hot*, and similar ones. I have argued that they are non-relational and therefore they are pure. But the properties of *instantiating the universal redness*, *having this red trope*, and *being a member of that class* are such that having them consists in being related in a certain way to a determinate entity (a certain universal, trope, and class, respectively), and therefore they are impure. But such properties, even if it is necessary that having the property of *being red* is necessarily connected to having one of them, are not the property of *being red*. Indeed, even the property of *satisfying the predicable condition of being red* is relational and impure, and it is different from the property of *being red*. For when one predicates of something that it is red one predicates a non-relational condition of that object, not a condition that consists in being related to any universal, trope, class, or even the predicable condition one is predicating of it! The point is, thus, that what the condition of *being red* requires of any object for it to satisfy it is that it is red, not that it is related in any way to the condition itself, even if it is necessary that any red object is related in a certain way to the condition of *being red*.

There is another possible objection to the point that all non-relational properties are pure. Isn't the property of *being Aristotle* both non-relational and impure? No, there is no property of *being Aristotle*, for there is no predicate 'is Aristotle', since Aristotle is not a condition that can be predicated of an object. When we say things like 'That man is Aristotle', the 'is' is not the 'is' of predication, but the 'is' of identity. So what we are predicating on those occasions is being identical with Aristotle—the predicate in question is 'is identical with Aristotle'. Thus, there *is* a property of *being identical with Aristotle*, but it is a *relational* impure property.

How about properties like *pegasizes* or *being the thing that pegasizes*? They are impure properties since they are properties of identity. That is, the property of *being the thing that pegasizes* is the property of *being identical with Pegasus*. Indeed, when Quine introduced descriptions like 'the thing that pegasizes', he did so to guarantee that non-descriptive names, like 'Pegasus', could be brought under the umbrella of Russell's theory of descriptions, and he saw such descriptions as the linguistic expression of properties like *being Pegasus* (Quine 2000: 7). Given my point in the previous paragraph, he should have considered them the linguistic expression of properties like *being identical with Pegasus*. In any case, what he meant was that the expression 'is the thing that pegasizes' predicates the property of *being identical with Pegasus*. But since what that expression predicates is the property of *being the thing that pegasizes*, it follows that this property is the property of *being identical with Pegasus*, and therefore the property of *being the thing that pegasizes*, and similar ones, are impure properties.

Finally, note that often impure properties are thought of as *non-qualitative* properties (Adams 1979, Sider 1996, and Cowling 2015 are three examples among many others). But impure properties are qualitative and, since they describe objects, they represent *how* objects are. Nevertheless impure properties contain a non-qualitative element, since what partially determines whether something has an impure property is *which* object or objects it is related to in a certain way. This non-qualitative element does not render them non-qualitative, it only renders them impurely qualitative.

One might wonder in which sense the property of *being identical with Aristotle* describes how a thing is. That is, how is a property like *being identical with Aristotle* qualitative, if only impurely so? The qualitative element in such a property is given by the relation of identity. Such a property describes an object as maintaining a certain relation, namely the relation of identity, with Aristotle.[9]

1.9 Let me introduce another distinction between properties, that between intrinsic and extrinsic properties. This distinction does not coincide with the distinction between non-relational and relational properties, since there are relational properties that are intrinsic. Here

---

[9] In the past I have used the phrase 'qualitative difference' to designate extra-numerical difference (Rodriguez-Pereyra 2006: 205). But such a phrase is confusing. For saying that two objects differ qualitatively means that they differ with respect to at least one property, whether a purely qualitative or an impurely qualitative property. But properties of identity are impurely qualitative properties and so objects that differ with respect to them thereby differ qualitatively. However, differing with respect to a property of identity is not differing extra-numerically.

are two examples: the properties of *being identical with Napoleon* and *having two legs*. These are clearly relational properties: having them consists in bearing the relation of *being identical with* to Napoleon and the relation of *having* to two legs. But they are also intrinsic properties, since having them consists in having a relation either to oneself or to one's parts but not to what I shall call, for lack of a better word, an external object, where an external object is one that is numerically different from a given object and also fails to be a part of it—that is, an external object is one that is neither a proper nor an improper part of a given object.[10] Thus, some intrinsic properties are relational, some intrinsic properties are impure, like *being identical with Napoleon*, and some intrinsic properties are pure, like *having two legs* and *being round*. Similarly, some extrinsic properties are pure and some are impure. *Being a brother* is, for instance, extrinsic and pure, while *being a brother of William James* is extrinsic and impure. Thus, the distinctions between pure and impure properties and between intrinsic and extrinsic properties cut across each other.

But let me say a bit more about how I understand intrinsic properties. I take intrinsic properties to be those such that having them does not consist in being related, or failing to be related, in any way to any external object or objects. Extrinsic properties are those such that having them consists, at least partly, in being related, or failing to be related, in some way to some external object or objects.

This applies to both pure and impure intrinsic properties. For instance, neither *being spherical* nor *being identical with Aristotle* are such that having them consists in being related in any way to any external object or objects. And *being lonely* is correctly classified as extrinsic since having this property consists in not coexisting with any external object (indeed, here we need to use the clause that makes reference to failing to be related to external objects, since being lonely does not consist in being related by non-coexistence to every external object—it just consists in failing to be related by coexistence to any external object).

---

[10] Note that although proper parts are not external to their wholes, wholes are external to their proper parts.

Now, there is no uncontested definition of intrinsic properties and my definition will not escape that fate.[11] For instance, it classifies the property of *being such that there is a cube* as intrinsic, since having it does not consist in being related in any way to any external object: if it did, a lonely cube would fail to have the property of *being such that there is a cube*. And, clearly, having it does not consist in failing to be related in any way to any external object. However, this property is widely thought to be extrinsic (see, for instance, Langton and Lewis 2001: 354 and Marshall and Parsons 2001: 349). But given the purposes of this book I cannot here defend my definition of intrinsic properties. Nevertheless, I should point out that my definition of intrinsic properties classifies correctly the clear and central cases of the intrinsic/extrinsic distinction and it accommodates the main platitude about intrinsic properties, namely that having an intrinsic property is independent of how the rest of the world is (for a list of many statements of this platitude by different people see Eddon 2011: 315, fn. 2). It accommodates the platitude in so far as having an intrinsic property does not consist in being related, or failing to be related, in any way to any external object or objects, and in that sense having an intrinsic property is independent of how the rest of the world is.

It has been argued that the basic distinction is not that between intrinsic and extrinsic properties but that between having a property intrinsically and having a property extrinsically, and that one and the same property can be had both intrinsically and extrinsically (Bader 2013: 554; see also Figdor 2008). This raises the possibility that two objects may share all their NT-properties but differ in that one of them has a certain property intrinsically and the other one has it extrinsically. If so, shouldn't PII quantify not only over NT-properties but also on modes of having them? That complication is unnecessary. Given my abundant conception of properties, if an object has a property F intrinsically and another one has it extrinsically, the former has the property of *having F intrinsically* and the latter has the property of *having F*

---

[11] My definition shares the spirit of Francescotti's (1999), but I do not agree with many of the details of his exact definition. For a taste of the difficulties involved in defining intrinsic properties see, among many others, Lewis 1983a, Langton and Lewis 1998, Sider 1996 and 2001, Francescotti 1999, and Marshall and Weatherson 2018.

extrinsically. Thus, the difference between these objects in relation to how they have property F can be captured in terms of properties.

In the next chapter I shall use the distinctions between pure and impure properties, and between intrinsic and extrinsic properties, to introduce different versions of PII. And I shall use my conception of intrinsic properties in Chapter 4 when arguing that Black's world is possible.

2
Trivializing Properties and NT-properties

2.1 Since PII quantifies over NT-properties, its content is not clear unless NT-properties are precisely defined. My task in this chapter is to provide such a definition. My method will be to define trivializing properties first and then define NT-properties negatively as those which are not trivializing properties.

Defining trivializing properties is a task very rarely attempted and, as far as I know, the most developed attempts are Katz (1983) and Rodriguez-Pereyra (2006). What I shall say here is based on my article from 2006, but it goes beyond it in important ways. One such way is that now I do not take principle (1) to be a formulation of PII, as I made clear in the previous chapter. Thus, my goal in this chapter is not to characterize the weakest non-trivial version of PII, but simply to characterize the weakest version of PII. Another difference is that, unlike in 2006, I shall now provide two definitions of trivializing properties, and neither will be identical to that from the 2006 paper, although both will be clearly related to it. There is also a lot of material from that article that will not find space here. In particular, I shall not reproduce my criticism of Katz's proposal. Instead, I shall concentrate on my positive account.

As we saw in the previous chapter, properties of identity, properties like *being identical with Aristotle* and *being identical with Napoleon* trivialize principle (1), the principle that necessarily, no two objects share all their properties. Indeed, properties of identity are the paradigmatic trivializing properties.[1]

---

[1] The trivializing nature of properties of identity is recognized, among many others, in Black 1952: 155, Ayer 1954: 29, O'Connor 1954: 103–4, Adams 1979: 11, Katz 1983: 37–8, Legenhausen 1989: 626, Della Rocca 2005: 481, Ladyman and Bigaj 2010: 118.

Let us get more clear about what properties of identity are. They are properties of being identical with a certain object in particular. This can be further clarified by resorting to the property abstraction lambda-operator. Thus, properties of identity are those such that in their lambda-expression the open sentence from which the lambda-operator binds a variable consists only of an identity sign flanked by an individual variable and an individual constant (I shall take proper names and other referential devices like demonstratives, referential uses of definite descriptions, and even free variables under an assignment to play the role of individual constants). So properties like *being identical with a* and *being identical with Aristotle* are properties of identity because their lambda-expressions are, respectively, '$(\lambda x)(x = a)$' and '$(\lambda x)(x = \text{Aristotle})$'. But the properties of *being identical with something* and *being self-identical* are not properties of identity. Their lambda-expressions, '$(\lambda x)(\text{there is a } y \text{ such that } x = y)$' and '$(\lambda x)(x = x)$', do not satisfy the characterization of properties of identity.

It should be clear that properties of identity are trivializing properties. Indeed, one could establish principle (1) by arguing as follows. Take any object a. It must have the property of *being identical with a*. If any object b shares all the properties with a, b must also have the property of *being identical with a*. But then a and b are numerically identical. Therefore, a cannot share all its properties with any other objects.[2]

This generalizes. For, necessarily, every object must have one such property of identity, and no two objects can possibly share such properties. Therefore, it is necessarily the case that no two objects share all their properties, which is what principle (1) asserts. But using properties of identity in this way to establish principle (1) trivializes it, for to establish it on the basis that no two numerically different objects can share their properties of identity is to establish it on the basis that no two numerically different objects can be numerically identical, which is a triviality.

If properties of identity were the only trivializing properties, the problem of defining them would be easy. But although properties of

---

[2] Versions of this argument appear in many places; see, for instance, Whitehead and Russell 1925: 57; Brody 1980: 9, Katz 1983: 37, and Legenhausen 1989: 626.

identity are the paradigmatic trivializing properties, there are other trivializing properties. Consider for instance conjunctive properties one of whose conjuncts is a property of identity, like the property of *being identical with a and being green*. Someone could argue for principle (1) as follows. Take a green object *a*. Such an object must have the property of *being identical with a and being green*. If object *b* shares all the properties with *a*, *b* must also have the property of *being identical with a and being green*. But then *a* and *b* are numerically identical. Therefore, *a* does not share all its properties with any other object. Of course, not every object is green. But every object must have some conjunctive property one of whose conjuncts is a property of identity and the other is some other property. And such properties cannot be shared by numerically different objects, since sharing them entails sharing a property of identity, which numerically different objects cannot do. Thus, such considerations generalize to establish principle (1). But they render it trivial, for establishing it on the basis that objects cannot share their properties of identity is to establish it on the basis that no two numerically different objects can be numerically identical, which is a triviality. Thus, conjunctive properties one of whose conjuncts are properties of identity are trivializing properties and should therefore be excluded from the domain of quantification of PII.

Consider a property like *being numerically different from a*. This property is the complement of a property of identity, the property of *being identical with a*. I shall call any complement of a property of identity a *property of difference*. Now, although properties of difference can be shared by numerically different objects, no numerically different objects can share all their properties of difference since, being numerically different, one will have a property of difference that the other one lacks. Take, for instance, object *a*. That object must lack the property of *being numerically different from a*. But a numerically different object *b* must have the property of *being numerically different from a*, so there is a property that *a* and *b* cannot share. This clearly generalizes since no two numerically different objects can share all their properties of difference— otherwise, they would be numerically identical. But it is not possible that two numerically different objects are numerically identical. So one can use properties of difference to establish principle (1), according to which, necessarily, no two objects share all their properties. But using properties of difference to establish principle (1) trivializes it, for it establishes it on the basis that no two numerically different objects can be numerically identical, which is a triviality. Thus, properties of difference are trivializing properties.

Sometimes it is said that trivializing properties are those which are unshareable. But properties of difference show the error of such a conception, since they are trivializing but shareable: both you and I share the property of *being numerically different from Aristotle*. A fortiori, properties of difference are also such that they do not necessarily apply to one object only. Thus, it is wrong to think that what makes a property trivializing is that it necessarily applies to one object only. We shall soon see not only that there are other shareable trivializing properties, but also that there are unshareable NT-properties. Thus unshareablity is neither necessary nor sufficient for being trivializing.

Consider now disjunctive properties one of whose disjuncts is a property of difference. One example is the property of *being numerically different from a or not being green*—the complement of the property of *being identical with a and being green*. Take two numerically different objects, *a* and *b*. Object *b*, and every object other than *a*, will have both the property of *being numerically different from a or not being green* and the property of *being numerically different from a or being green*, but *a* can have only one of these. Thus there will be a disjunctive property, one of whose disjuncts is a property of difference, such that they will not share it; otherwise *a* and *b* will share all their properties of difference. This is, of course, perfectly generalizable for any two numerically different objects. But using disjunctive properties one of whose disjuncts is a property of difference to establish principle (1) trivializes it, for such a principle would then be established on the basis that no two numerically different objects can share all their properties of difference, which is establishing it on the basis that no two numerically different objects can be numerically identical, which is a triviality.

It might be thought that only one of *being numerically different from a or not being green* and *being numerically different from a or being green* is trivializing. For instance, if *a* is green, it does not differ from any other object with respect to the property of *being numerically different from a or being green*, and so the trivializing property must be the property of *being numerically different from a or not being green* (similarly, if *a* is not green, it does not differ from any other object with respect to the property of *being numerically different from a or not being green*, and so the trivializing property must be the property of *being numerically different from a or being green*).[3] True, but such properties can be used to prove that no two objects can share all their properties in a very general and abstract way, without presupposing anything about the colour of any object. Indeed, that is how I just showed that they could be used to establish that no two objects can share all their properties: all I assumed is that there are two objects, *a* and *b*, and that, consequently, since *b* must have both such properties but *a* can have only one of them, no matter what other properties *a* and *b* have, they cannot share all their properties. Because of this there is no reason to count either of them as trivializing that does not apply to the other, and so it would be arbitrary to count only one of them as trivializing. Thus we should count both of them, and any properties like them, as trivializing properties.

Disjunctive properties one of whose disjuncts is a property of identity are also trivializing. Consider the properties of *being identical with a or being green* and *being identical with a or not being green*. Every object must have at least one of them. But only object *a* can, and must, have both of them—any other object having both of them would share a property of identity with *a*, which is impossible. Thus, such properties show that no object can share all its properties with *a*. Clearly, this reasoning generalizes to establish principle (1). But using disjunctive properties one of whose disjuncts is a property of identity to establish principle (1) trivializes it, since it establishes it on the basis that no two numerically different objects can share their properties of identity, which is establishing it on the basis that no two numerically different objects can be numerically identical, which is a triviality.

It might be thought that only one of *being identical with a or being green* and *being identical with a or not being green* is trivializing. For instance, if *b* is green, it does not differ from *a* with respect to the property of *being identical with a or being green*, and so the trivializing property

---

[3] I owe this point to Michael Della Rocca.

must be the property of *being identical with a or not being green* (similarly, if *b* is not green, it does not differ from *a* with respect to the property of *being identical with a or not being green*, and so the trivializing property must be the property of *being identical with a or being green*). True, but such properties can be used to prove that no two objects can share all their properties in a very general and abstract way, without presupposing anything about the colour of any object. Indeed, that is how I just showed that they could be used to establish that no two objects can share all their properties: all I assumed is that there are two objects, *a* and *b*, and that, consequently, since *b* can have at most one of those properties but *a* can and must have both of them, no matter what other properties *a* and *b* have, they cannot share all their properties. Because of this there is no reason to count either of them as trivializing that does not apply to the other, and so it would be arbitrary to count only one of them as trivializing. Thus we should count both of them, and any properties like them, as trivializing properties.

Similarly with properties like *being numerically different from a and not being green* and *being numerically different from a and being green*. Every object must lack at least one of them. But only object *a* can, and must, lack both of them. Thus, such properties show that no object can share all its properties with *a*: if any other object shared all its properties with *a*, it would lack both such properties and so, being either green or not green, it would lack the property of *being numerically different from a* and instead have the property of *being identical with a*, but it is impossible for *a* to share its property of identity with a numerically different object. Furthemore, every object must lack a pair of properties like those two. So such properties guarantee that no two objects can share all their properties, which is what principle (1) states. But they also trivialize it, since establishing it on the basis that no two numerically different objects can share a property of identity is establishing it on the basis that no two numerically different objects can be numerically identical, which is a triviality.

It might be thought that only one of *being numerically different from a and not being green* and *being numerically different from a and being green* is trivializing. For instance, if *b* is green, it does not differ from *a* with respect to the property of *being numerically different from a and not being green*, and so the trivializing property must be the property of *being numerically different from a and being green* (similarly, if *b* is not green, it does not differ from *a* with respect to the property of *being numerically different from a and being green*, and so the trivializing property must be the property of *being numerically different from a and not being green*). True, but such properties can be used to prove that no two objects can share all their properties in a very general and abstract way, without presupposing anything about the colour of any object. Indeed, that is how I just showed that they could be used to establish that no two objects can share all their properties: all I assumed is that there are two objects, *a* and *b*, and that, consequently, since *b* can lack at most one of those properties but *a* can and must lack both of them, no matter what other properties *a* and *b* have, they cannot share all their properties. Because of this there is no reason to count either of them as trivializing that does not apply to the other, and so it would be arbitrary to count only one of them as trivializing. Thus we should count both of them, and any properties like them, as trivializing properties.

There are many more trivializing properties. Just another couple of examples: the property of *(being identical with a or not being green) and being green*, which is logically equivalent to the property of *being identical with a and being green*, and the property of *being a member of {a}*. Consider the latter property. Every object must have a property like that, since every object must be a member of its singleton. But objects cannot share those properties, for the identity of objects determines which singletons they are members of. Thus, what is doing the work here are properties of identity, as we shall see more clearly in Section 2.5. Such properties are, then, trivializing properties.

2.2 Since properties of identity are the paradigmatic trivializing properties, it seems plausible that trivializing properties are those related to properties of identity in a special way. One might try defining trivializing properties in terms of a notion of property entailment, where a property F entails a property G if and only if, necessarily, everything having F has G. This is the standard way of conceiving of property entailment (see, for instance, Katz 1983: 44, fn 4, Lewis 1983a: 199, and Carnap 1988: 17). Of course, it will not do to say that trivializing properties are those which entail a property of identity, since not all trivializing properties entail a property of identity: for instance, there is no property of identity that is entailed by the property of *being numerically different from Aristotle*.[4]

One might, however, complicate the definition and say that trivializing properties are those such that they or their complements entail a property of identity. Such a definition is satisfied by properties of difference. But it is not satisfied by all trivializing properties. Consider, for instance, the property of *being identical with a or being green*. No property of identity is entailed by it or its complement. (For a criticism of a further definition of trivializing properties in terms of entailment see Rodriguez-Pereyra 2006: 210).

Recently, David Wörner (2021) has proposed a relativized definition of trivializing properties according to which a property F is trivializing with regard to x and y if and only if x and y differ with regard to F partly in virtue of the fact that x is numerically distinct from y. But what if every negative fact is grounded only in positive ones? That is a view that cannot be ruled out on the basis of an account of trivializing properties. And yet, assuming the evident fact that some properties are trivializing, Wörner's account rules out such a view. For x and y's differing with respect to any property is a negative fact and so, on the view that every negative fact is grounded only in positive ones, it cannot be partly grounded in the fact that x and y are numerically distinct, since this is also a negative fact. Therefore, on such a view, Wörner's account entails that there are no trivializing properties, which is clearly false (and would be rejected by Wörner himself). True, the view that every negative fact is grounded only in positive ones might well be false. But the point is that, since what trivializing properties are is something completely unrelated to whether and how negative facts are grounded, accounts of trivializing properties should be neutral about that view. Therefore, that an account of trivializing properties rules out the view that negative facts are

---

[4] But doesn't the property of *being numerically different from a* entail the property of *being identical with b or being identical with c or being identical with d...*, where $b, c, d...$ are all the objects numerically different from a? Not unless nothing different from $a, b, c, d...$ can coexist with a. But even if this is the case, the property of *being identical with b or being identical with c or being identical with d...*, is not a property of identity but a disjunctive property whose disjuncts are properties of identity.

grounded only in positive ones, is an indication that the account in question is defective.[5]

Alternatively, one might try to define trivializing properties in terms of property containment. To make this fly one needs to make the notion of containment precise and one constraint is that properties of identity contain themselves so that they can be counted as trivializing properties (for criticism of some attempts to define trivializing properties in terms of containment, see Rodriguez-Pereyra 2006: 211–14). In Section 2.6 I shall propose two definitions of trivializing properties, one of which will be in terms of the notion of property containment. A virtue of my definitions is that they are based on an explanation of what makes trivializing properties trivializing. This is important, for if we do not know why properties of identity are trivializing, how can we know that any proposed definition of trivializing properties is even extensionally correct? Thus, what we primarily need is an explanation of why trivializing properties are trivializing. Once we have such an explanation we can easily define trivializing properties: trivializing properties are those which have the features in question, and we shall know such a definition to be extensionally correct. And on the basis of that definition one can understand NT-properties as those lacking the definional features of trivializing properties.

2.3 There are two questions I need to answer: what is it to be a trivializing property?, and what makes trivializing properties trivializing? The answer to the first question describes the character of trivializing properties. The answer to the second question tells us in virtue of what trivializing properties have the trivializing character.

Since every object must have a property of identity, and no two objects can share their properties of identity, properties of identity guarantee that principle (1) is true. If one establishes principle (1) on this basis, all one has proved is that objects must differ with respect to their properties of identity. But differing with respect to a property of identity *is* differing numerically, and therefore merely establishing a difference with respect to properties of identity does not establish an extra-numerical

---

[5] Thanks to Tien-Chun Lo for discussion of this point.

difference. Thus, if one establishes principle (1) on the basis that objects must differ with respect to their properties of identity, one has only established that necessarily, any objects that differ numerically differ numerically, which is a triviality.

What makes properties of identity have this feature? That, given the properties they are, differing with respect to them is differing numerically. Indeed, two objects differ with respect to a property of identity if and only if one of them has it and the other one lacks it, that is, if and only if one of the objects is numerically identical with a certain object and the other one is numerically different from that object. But differing in that way is differing numerically. Therefore, since differing extra-numerically is differing more than numerically, merely establishing a difference with respect to properties of identity does not establish an extra-numerical difference.

It is important to distinguish clearly between the trivializing character of properties of identity, and what makes them have that character. The trivializing character of properties of identity consists in that merely establishing that two objects differ with respect to them establishes only a numerical difference between them. What makes properties of identity have that character is that differing with respect to them is differing numerically.

The case of properties of difference is exactly similar. Their trivializing character consists in that merely establishing that two objects differ with respect to their properties of difference is establishing no more than that they are numerically different. And what makes them have the trivializing character is that differing with respect to them is differing numerically. Indeed, for two objects to differ with respect to a property of difference is for one of them to have the property of difference in question and for the other object to lack it. But lacking a property of difference is having the corresponding property of identity, and having a property of difference is lacking the corresponding property of identity. Therefore, differing with respect to properties of difference is differing with respect to properties of identity, and therefore differing with respect to properties of difference is differing numerically.

It is important to note that the trivializing character of properties of identity and difference consists in that merely establishing a difference with respect to them only establishes that the objects in question are numerically different—not that the objects in question differ only with respect to properties of identity and difference. The latter is not true. For instance, if *a* and *b* differ with respect to properties of identity and difference then they differ with respect to conjunctive properties that have their properties of identity and difference as conjuncts. Such conjunctive properties are trivializing properties, but they are not properties of identity or difference.

Even if PII is true, and differing with respect to properties of identity and difference entails differing with respect to NT-properties, properties of identity and difference are still trivializing properties. Consider, for instance, pure individual essences. If a property is a pure individual essence, it is a pure property that is a metaphysically necessary and sufficient condition for being a certain object in particular. Thus, suppose that, necessarily, all objects have pure individual essences, and that Aristotle's pure individual essence is the property of *being the greatest philosopher ever*. Pure individual essences, like all pure properties, are NT-properties, since differing with respect to them is differing extra-numerically, and therefore if objects must have pure individual essences, PII is true. So, if objects must have pure individual essences, differing with respect to properties of identity necessitates differing with respect to pure individual essences, and a difference with respect to pure individual essences is an extra-numerical difference. But even then if one merely establishes that numerically different objects must differ with respect to properties of identity or difference one has only established the triviality that numerically different objects must differ numerically. For whatever differences necessarily accompany differences with respect to properties of identity and difference, differing with respect to properties of identity and difference is differing numerically. And since differing with respect to them is differing numerically, when one merely establishes that numerically different objects must differ with respect to properties of identity or difference one only establishes the triviality that numerically different objects must differ numerically. Thus, properties of identity and difference are still trivializing properties even if differing with respect to them necessitates differing with respect to NT-properties.

2.4 The trivializing character is common to all and only trivializing properties. Every property such that merely establishing a difference with respect to it only establishes that the objects in question are numerically different is a trivializing property. And every property such that merely establishing a difference with respect to it establishes an extra-numerical difference is an NT-property.

But although the trivializing character is common to all and only trivializing properties, only in the case of properties of identity and difference what makes them trivializing is that differing with respect to them is differing numerically. Indeed, to give just two examples, to differ with respect to the property of *being identical with a or being green* is not to differ numerically, since it is to differ with respect to at least one of its disjuncts, and so two objects might differ with respect to it by differing with respect to *being green*. Similarly, to differ with respect to the property of *being identical with a and being green* is not to differ numerically, since it is to differ with respect to at least one of its conjuncts, and although whenever two objects differ with respect to it they must differ with respect to the property of *being identical with a*, they might also differ with respect to the property of *being green*.

Now, even if differing with respect to property F is not differing numerically, it is sometimes the case that merely establishing that numerically different objects must differ with respect to property F only establishes that numerically different objects must differ numerically. This happens when differing with respect to F is consistent with differing only numerically. For if differing with respect to F is consistent with differing only numerically, then merely establishing that numerically different objects must differ with respect to F only establishes the triviality that numerically different objects must differ numerically.

The relevant sense of consistency at play here is the following: differing with respect to F is consistent with differing only numerically if and only if it is not a requirement *of the property F itself* that any objects differing with respect to it must differ extra-numerically. When it is not a requirement of the property F itself that any objects differing with respect to it must differ extra-numerically, I shall say that differing with respect to F does not require differing extra-numerically.

It is important to emphasize that, for a property to be trivializing, it must not be a requirement *of the property in question* that any objects differing with respect to it must differ extra-numerically. For, as we saw at the end of the last section, it might be that a difference with respect to properties of identity or difference is necessarily accompanied by a difference with respect to pure individual essences or, indeed, any extra-numerical difference. But even if this is the case, this does not stop properties of identity and difference from being trivializing, since from the mere fact that two objects differ with respect to them one cannot infer—and therefore one cannot establish—that those objects differ extra-numerically. And this is because it is not a requirement of the properties of identity and difference themselves that any objects differing with respect to them must differ extra-numerically. Of course, if one knows that objects have pure individual essences, or that any two numerically different objects must differ extra-numerically in some way or another, one is in a position to infer that two objects differ extra-numerically from the fact that they differ numerically. But such an inference would involve something more than a mere appeal to the fact that the objects in question differ with respect to properties of identity and difference: it would also involve an appeal to the fact that objects have pure individual essences, or that any two objects must differ extra-numerically in some way or another. And these are not trivial facts.

Let us confirm this account of what makes properties other than properties of identity and difference trivializing properties. Differing with respect to the property of *being identical with a and being green* does not require differing extra-numerically, since two objects could differ with respect to it by simply differing with respect to the property of *being identical with a*. But then, if two objects could differ with respect to the conjunctive property by simply differing with respect to a property of identity, merely establishing a difference with respect to the property of *being identical with a and being green* only establishes a numerical difference. Therefore, the property of *being identical with a and being green* is a trivializing property. This clearly generalizes to all conjunctive properties one of whose conjuncts is a property of identity.

Of course, any two objects differing with respect to *being identical with a and being green* must differ with respect to the property of *being identical with a*, since to differ with respect to the conjunctive property one of the objects must have both conjuncts and the other one must lack at least one, but no two objects can share a property of identity. But that objects differing with respect to such conjunctive property must differ with respect to a property of identity is irrelevant: any objects differing with respect to any property must differ with respect to properties of identity, since no two objects can differ with respect to any property without being numerically different. What matters is not that differing with respect to the conjunctive property requires differing with respect to a property of identity but that it does not require differing extra-numerically.

The same applies to conjunctive properties containing properties of difference, like the property of *being numerically different from a and being green*. Two objects could differ with respect to it by simply differing with respect to the property of *being numerically different from a*, and therefore differing with respect to the property of *being numerically different from a and being green* does not require differing extra-numerically. But then merely establishing a difference with respect to the property of *being numerically different from a and being green* only establishes a numerical difference. Therefore, the property of *being numerically different from a and being green* is a trivializing property. This clearly generalizes to all conjunctive properties one of whose conjuncts is a property of difference.

Again, the same applies to disjunctive properties having a property of identity as disjunct, like the property of *being identical with a or being green*. Two objects could differ with respect to the disjunctive property by simply differing with respect to the property of *being identical with a* (that is how two red objects will differ with respect to it). Thus, differing with respect to the disjunctive property does not require differing extra-numerically. But then merely establishing a difference with respect to the property of *being identical with a or being green* only establishes a numerical difference. Therefore, the property of *being identical with a or being green* is a trivializing property. This clearly generalizes to all disjunctive properties one of whose disjuncts is a property of identity. Needless to say, the same story applies to disjunctive properties one of whose disjuncts is a property of difference.

It should be clear now *why* properties like *being green*, *being square*, and *being hot* are NT-properties. Differing with respect to them requires differing extra-numerically because, given the properties they are, differing with respect to them requires differing not only numerically but also with respect to colour, shape, and temperature.

Indeed, it should be clear that all pure properties are NT-properties. For pure properties do not consist in being related in any way to any object in particular, and therefore, given what they are, differing with respect to them must require a difference other than a mere numerical difference, that is, it must require an extra-numerical difference.

For similar reasons impure properties like *being a parent of a, loving b, being close to c,* and *being in the same place as d*, are NT-properties. Given what these properties are, differing with respect to them requires differing extra-numerically: it requires, respectively, differing with respect to parenting *a*, loving *b*, being close to *c*, and being in the same place as *d*. Let *e* be a parent of *a*. Even if origin is essential, and so *e* cannot fail to be a parent of *a* provided *a* exists, there is more to being a parent of *a* than being such that *a* exists and being identical with *e*. The extra is all that is involved in parenting *a*. Similarly in the other cases.

An interesting case is the property of *having all parts in common with a*. This is trivializing because among the parts of *a* is its improper part, namely *a* itself. So this property leaves open the possibility that *a* and *b* differ only with respect to their improper parts, in which case they differ only numerically. Some people think that no two objects can share all their proper parts. If they are right, the property *having all proper parts in common with a* cannot be shared. But this does not make it trivializing. Differing with respect to that property is differing more than numerically: it is differing with respect to which objects are one's proper parts.

This also explains why superlative properties, properties like *being the tallest man* and *being the widest river*, are NT-properties. Differing with respect to *being the tallest man* is more than differing numerically: it is also differing with respect to the difference in height in relation to other men. Superlative properties are interesting because they show that being unshareable does not entail being trivializing.

It should also be noted that there are some properties containing properties of identity that are NT-properties, namely those properties that contain properties of identity but are *logically* equivalent to NT-properties, like the property of *being green and (being identical with a or being numerically different from a)*. What this property requires is that objects differing with respect to it must differ with respect to at least one of the conjuncts—that is, of any two objects differing with respect to it one of them must lack at least one of the conjuncts. But it is a requirement of the second conjunct, the property of *being identical with a or being numerically different from a*, that of any two objects differing with respect to it one of them must lack both disjuncts. But since each disjunct is the complement of the other, nothing can lack both of them and therefore no two objects can differ with respect to it. But then it is a requirement of the property of *being green and (being identical with a or being numerically different from a)* that any two objects differing with respect to it must differ with respect to the first conjunct, the property of *being green*. And therefore it is a requirement of the property of *being green and (being identical with a or being numerically different from a)* that any two objects differing with respect to it must differ extra-numerically, namely with respect to colour. Thus, establishing a difference with respect to that property is establishing an extra-numerical difference, and therefore the property of *being green and (being identical with a or being numerically different from a)* is an NT-property. The account generalizes to other properties that contain properties of identity but are logically equivalent to NT-properties.

Some philosophers have claimed that predicates in which identity occurs express trivializing properties (see, for instance, Muller 2015: 211; cf. Dorato and Morganti 2013: 594). But, as we have just seen in relation to the property of *being green and (being identical with a or being numerically different from a)*, this is not the case.

2.5 So far, so good. But how about the property of *being a member of {a}*? This is a trivializing property but it does not seem to satisfy my characterization of trivializing properties. For differing with respect to it seems to require not only differing with respect to being identical with *a*, but also differing with respect to being a member of {a}, and so it seems that differing with respect to it entails differing extra-numerically. On the other hand, however mysterious the singleton membership relation is, it seems that differing with respect to being a member of a singleton is no more than differing numerically. How can this be?

This is because, if sets exist, the identity of set-members fixes what sets they belong to. And this is, in turn, because given a set S with certain objects as members, there is no more to being a member of S than being one of those objects. So, given $\{a\}$, there is no more to being a member of $\{a\}$ than being a, that is, being identical with a. Thus, the property of *being a member of $\{a\}$* is the property of *being such that $\{a\}$ exists and being identical with a*.[6]

Since the property of *being such that $\{a\}$ exists and being identical with a* is a conjunctive property one of whose conjuncts is the property of *being identical with a*, differing with respect to the property of *being a member of $\{a\}$* does not require differing extra-numerically. Thus the property of *being a member of $\{a\}$* is a trivializing property. Indeed, since whenever two things differ with respect to being a member of $\{a\}$ both of them are such that $\{a\}$ exists, any objects differing with respect to *being a member of $\{a\}$* must always differ with respect to the property of *being identical with a*.

Let me now briefly discuss some interesting aspects of my account of the property of *being a member of $\{a\}$*. It is clear that a belongs to $\{a\}$ in virtue of being identical with a rather than being identical with a in virtue of belonging to $\{a\}$. My proposed account of the property of *being a member of $\{a\}$* nicely explains why this is so: being a member of $\{a\}$ consists in satisfying two conditions, one of which is being identical with a, but being identical with a does not consist in being a member of $\{a\}$, not even partially.

My account of the property of *being a member of $\{a\}$* also goes some way to dispel the mystery associated with the singleton membership relation. According to David Lewis, the relation of singleton to member holds in virtue of qualities and relations of which we have no conception whatsoever. That is, we do not clearly understand what it is for a singleton to have a member (Lewis 1991: 35). If my account of the property of *being a member of $\{a\}$* is right then we have a conception of the relations

---

[6] This goes against the *reductive* definition of identity in set-theoretic terms, according to which identity between objects consists in their being members of the same sets. For discussion of defining identity in terms of sets see, for instance, Burgess 2012: 95–6.

in virtue of which an object is a member of a singleton: these are existence and identity.

Nevertheless, there is a sense in which Lewis is right that the singleton membership relation is mysterious. Singletons are atoms and the connection with their members is primitive and thereby in some sense mysterious and opaque. So, we do not know in virtue of what the property of *being a member of {a}* is the property of *being such that {a} exists and being identical with a* rather than the property of *being such that {a} exists and being identical with b*. After all, if {a} exists, b has the property of *being such that {a} exists and being identical with b*; but this does not make it a member of {a}.

But this is a mystery that we should expect. For there is nothing in virtue of which {a} is the singleton of *a* as opposed to the singleton of *b*: it just is (here I go beyond Lewis, who seems to think that there may be something in virtue of which singletons have their members). So there is nothing in virtue of which the property of *being a member of {a}* is the property of *being such that {a} exists and being identical with a* rather than the property of *being such that {a} exists and being identical with b*: it just is one rather than the other.

Wörner (2021) thinks that properties like *grounding the existence of {a}* are trivializing properties. He gives no reason for his claim. But here is a reason for the opposite claim, the claim that properties like *grounding the existence of {a}* are NT-properties. For differing with respect to such a property is differing more than merely numerically, since it is differing with respect to which singleton set one grounds. But isn't then differing with respect to being a member of {a} differing more than merely numerically, since it is differing with respect to which singleton set one is a member of? There is a difference between the two properties. Since sets are extensional entities there is no more to belonging to a set than being identical with one of its members, and this is why I take the property of *being a member of {a}* to be the property of *being such that {a} exists and being identical with a*. But there is more to grounding a singleton set, or grounding any entity for that matter, than being identical with a certain entity, *even if a grounds {a} in virtue of being identical with a*. Indeed, in general, if an entity *x* grounds an entity *y* in virtue of *x*'s being F, one needs to distinguish between *x*'s being F and *x*'s grounding *y*. Thus, differing with respect to grounding the existence of {*a*} requires more than a mere numerical difference even if *a* grounds the existence of {*a*} in virtue of being identical with *a*. Therefore, *grounding the existence of {a}*, and similar properties, are NT-properties.

2.6 We have seen that what makes a property trivializing is that differing with respect to it does not require differing extra-numerically, and this happens when differing with respect to it is differing numerically and when objects could differ with respect to it by simply differing numerically, for in both cases merely establishing a difference with respect to that property only establishes a numerical difference. But there are properties with respect to which it is impossible to differ, since either everything must have them, or everything must lack them. For instance, properties like *being identical with a or being numerically different from a, being an object, existing, being identical with a and being numerically different from a*, and *being square and round*. Are such properties trivializing properties or NT-properties?

Although nothing much depends on this issue, I shall argue that they are NT-properties. The point is that trivializing properties are those that render principle (1) trivially true, but properties with respect to which no objects can differ cannot render principle (1) true, and therefore they cannot render it trivially true. For (1) states that, necessarily, no two objects share all their properties, which is equivalent to the claim that, necessarily, every two objects differ in at least one property. But properties over which no objects can differ cannot be used to establish such a claim. For these reasons I think properties that must be shared or lacked by every object should be counted as NT-properties.

But since PII states that, necessarily, no two objects share all their NT-properties, shouldn't those properties be counted as neither trivializing properties nor NT-properties? After all, someone might say, they cannot be used to establish PII either, since no two objects can differ with respect to them. But this reasoning is wrong. NT-properties are those that do not render (1) trivially true; whether or not they can be used to establish PII is irrelevant to their status as NT-properties.

Trivializing properties and NT-properties can be defined as follows:

D1. F is a trivializing property $=_{\text{def.}}$ (i) Possibly, two objects differ with respect to F, and (ii) Differing with respect to F does not require differing extra-numerically.

D2. F is an NT-property $=_{\text{def.}}$ F is not a trivializing property.

Is D1 extensionally correct? Yes. For, as I have argued, D1 is based on the features that make trivializing properties trivializing. *A fortiori*, D2, the definition of NT-properties is also extensionally correct.

Now, an alternative definition of trivializing properties can be given in terms of a simple notion of property containment. I am not in a position to give a full theory of property containment and, in any case, doing that would be unnecessary given my purposes. To give the alternative definition of trivializing properties all I need to do is to make relatively clear the idea of a property containing another one. To do this I shall resort to the structure of lambda-expressions.

Remember that to every meaningful predicate there corresponds a property. Predicates can be regimented as open sentences, which are sentences with one or more free variables, but since properties are monadic the open sentences corresponding to them have only one free variable. The lambda-expression of a property binds a variable from an open sentence—and sometimes, if the open sentence in question has other open sentences as parts, it binds its variable from a part of the main open sentence. Let us then say that a property F contains a property G if and only if F's lambda-expression binds its variable from an open sentence corresponding to a predicate expressing G. Thus the property of *being red and being round* contains the property of *being round*, since its lambda-expression binds a variable from an open sentence corresponding to a predicate expressing the property of *being round*: $(\lambda x)(x$ is red and x is round). Similarly, the property of *being identical with a or being green* contains the property of *being identical with a*, since its lambda-expression is: $(\lambda x)(x = a$ or x is green). Obviously, all properties contain themselves. We can now distinguish between *atomic* and *non-atomic* properties: atomic properties are those which contain only themselves, non-atomic properties are those which contain other properties as well. For instance, the properties of *being green* and *being identical with Aristotle* are atomic properties—indeed, all properties of identity are atomic properties. But the properties of *being red and being round*, *not being green*, *being coloured because of being green*, and *being identical with Aristotle and being green* are non-atomic properties. Note, also, that properties of difference are non-atomic, since they contain properties of identity: for instance, the property of *being numerically different from Aristotle* is the property of *not being identical with Aristotle*, and so it contains the property of *being identical with Aristotle*. Then one can define trivializing properties in terms of property containment as follows:

D3. F is a trivializing property =$_{\text{def.}}$ (i) Possibly, two objects differ with respect to F, (ii) F contains at least one property of identity, and (iii) If G is any atomic property contained in F, and differing with respect to F requires differing with respect to G, G is a property of identity.

It is important to note that sometimes the third clause of D3 will be vacuously met, since there are trivializing properties such that differing with respect to them does not require differing with respect to any particular atomic property they contain. For instance, differing with respect to the property of *being identical with a or being green* requires neither differing with respect to the property of *being identical with a* nor with respect to the property of *being green*, since differing with respect to the property of *being identical with a or being green* is compatible with differing with respect to either of its disjuncts.

D1 and D3 are equivalent in the sense that the properties satisfying one are exactly the same as the properties satisfying the other. Let us see this. Since the first clause of both definitions is the same, we can ignore that clause and concentrate only on the others. Given what I have said about properties differing with respect to which does not require differing extra-numerically, it is clear that any property satisfying D1 will satisfy the second clause of D3. But satisfaction of the second clause of D3 does not guarantee satisfaction of D1. Consider, for instance, the property of *being green and (being identical with a or being numerically different from a)*. This property satisfies the second clause of D3, since it contains the property of *being identical with a*. And yet differing with respect to it entails differing extra-numerically since, as we saw at the end of Section 2.4, differing with respect to it requires differing with respect to *being green*. The third clause in D3 fixes this problem. Indeed, the property of *being green and (being identical with a or being numerically different from a)* fails to satisfy the third clause in D3 and therefore it is an NT-property. In general, since differing with respect to a property of identity is differing numerically, any property non-vacuously satisfying the third clause of D3 will be such that differing with respect to it does not require differing extra-numerically. Therefore, properties satisfying D3 satisfy D1.

Now, properties satisfying D1 satisfy the third clause of D3 too. For take any property F satisfying D1. It must be either atomic or non-atomic. If it is atomic, then it is a property of identity, and since properties contain themselves, it will contain itself and so F will non-vacuously satisfy the third clause of D3. And if it is non-atomic, since it satisfies the second clause of D3, it contains at least one property of identity. Now, any atomic property G contained in F such that differing with respect to F requires differing with respect to G, must be a property of identity. Otherwise, differing with respect to F would require differing extra-numerically, and so F would not satisfy D1. And if there is no atomic property G contained in F such that differing with respect to F requires differing with respect to G, F vacuously satisfies the third clause of D3. Thus, in every case, any property satisfying D1 satisfies D3.

Thus D1 and D3 are equivalent in the sense that the properties satisfying one are exactly the same as the properties satisfying the other. Therefore, since D1 is extensionally correct, so is D3.

But why offer two definitions of trivializing properties? The philosophically fundamental one is D1, since it is based on what makes trivializing properties trivializing. D1 gives not only an extensionally correct characterization of trivializing properties but also a philosophical explanation of them. But D3 offers an easier criterion to decide whether a property is trivializing or not, since whether a property contains a property of identity is not a very difficult matter to determine.

For the sake of example, let us see how D3 classifies certain properties. It is obvious that it correctly classifies as trivializing the properties of *being identical with Aristotle, being numerically different from Aristotle, being green and being identical with Aristotle, being green or being numerically different from Aristotle,* and *(being green or being identical with Aristotle) and (not being green or being identical with Aristotle).* And D3 rightly fails to classify as trivializing properties such properties as *being green, being two meters away from a tower,* and *being a teacher of Aristotle.* This should be obvious in the first two cases, but it is also easy to see in the third: the property of *being a teacher of Aristotle* does not contain a property of identity, which can be easily seen by consulting its lambda-expression: (λx)(x is a teacher of Aristotle). This makes it clear, in case it was necessary, that the property of *being a teacher of Aristotle* is an atomic property and so it only contains itself.

It might be objected that a more perspicuous lambda-representation of the property is this: (λx)(x is such that there is a y such that y = Aristotle and x is a teacher of y). But this does not support the idea that such a property contains a property of identity, since the variable bound by the lambda-operator, 'x', does not occur in the open sentence corresponding to a predicate of identity ('y = Aristotle').

But isn't the following at least an equally perspicuous lambda-representation of the property in question: (λx)(x is such that there is a y such that y has the property (λy)(y = Aristotle) and x is a teacher of y)? But the variable bound by the lambda-operator at the front, 'x', does not occur in the open sentence corresponding to a predicate of identity ('y = Aristotle'). Thus, the property expressed by such lambda-representation does not contain a property of identity. Another way of making the point is the following: the condition of *being identical with Aristotle* does not have to be attached to the object satisfying the condition of *being such that there is someone identical with Aristotle and being a teacher of him*, which is manifested by the fact that these two conditions have different variables associated with them.

A very different case is, of course, the property of *being a teacher of Aristotle and being identical with Aristotle.* This property does contain a property of identity. But the property of *being a teacher of Aristotle and being identical with Aristotle* is not the property of *being a teacher of Aristotle.* This should be clear, since in predicating those conditions of one object one is predicating different conditions of it. Thus, the property of *being a teacher of Aristotle* does not contain a property of identity and D3 rightly fails to classify it as a trivializing property.[7]

2.7 We are now in a position to understand the formulation of PII:

PII: Necessarily, no two objects share all their NT-properties.

PII is evidently not a trivial principle, and yet it quantifies over impure properties. For, as we have seen, an impure property like *being a teacher of Aristotle* is an NT-property. And there are myriads of other impure NT-properties: *being next to the Eiffel Tower, corresponding with Leibniz, defeating Napoleon, visiting the Louvre, climbing Everest, having the same height as Aristotle*, and so on. This is interesting because it is often thought that quantifying over impure properties makes PII trivial.[8] But, as we have seen, it is only a subset of impure properties that are trivializing properties.

Note that although properties of identity are trivializing properties, and therefore PII does not quantify over them, the fact that PII quantifies over other impure properties shows that PII is compatible with primitive

---

[7] Ladyman, Linnebo, and Pettigrew (2012: 169) distinguish between identity-involving properties (those that involve the identity relation, but they need not be my properties of identity) and object-involving properties (those that appeal to particular objects—basically, my impure properties). Since trivializing properties contain properties of identity, and properties of identity are both identity-involving and object-involving, all trivializing properties are both identity-involving and object-involving properties. But not all identity-involving properties and not all object-involving properties are trivializing properties. Here are two examples: *being identical with something* is an identity-involving NT-property, and *being a teacher of Aristotle* is an object-involving NT-property.

[8] Among the many who are explicitly or implicitly committed to the idea that all impure properties are trivializing are Odegard 1964: 209, Robinson 2000: 164, Della Rocca 2005: 481–2, Ladyman and Bigaj 2010: 118, Shiver 2014: 907, Goodman 2015: 1, footnote 1, Muller 2015: 227, Wiggins 2016: 8, Shumener 2017: 4, and all those who think that properties corresponding to predicates featuring proper names are trivializing properties (of course, not all these authors use the term 'impure properties' or, if they do use it, they do not necessarily use it exactly in my sense, but all of them are committed to the idea that impure properties, in my sense, are trivializing properties). On the other hand, McTaggart seems to have been prepared to quantify over impure properties (although he did not know them under this name): 'Two things are not exactly similar if one of them is known to me and the other one is not.... For then one has, and one has not, the quality of being cognized by a *particular* person at a particular time.' (McTaggart 1921: 96; my emphasis). However, although he is aware that properties of identity and properties of difference should not be included in the domain of PII, McTaggart doesn't explain which impure properties are trivializing and which ones are not.

identity, namely the thesis that the numerical identity of objects is not grounded in—nor does it supervene upon—their purely qualitative properties. This is interesting, because it goes *contra* the received view (see, for instance, Adams 1979 and Della Rocca 2005). But since what PII requires is that numerical difference must be accompanied by extra-numerical difference, not that it must be accompanied by purely qualitative difference, the truth of PII is compatible with primitive identity in the sense that PII does not require purely qualitative difference for numerical difference. For all PII says, objects might differ only with respect to impure properties, and such properties involve the identity of other objects and even of themselves.

At this point some people might object that even if PII is not trivial, my enterprise in this book is futile. Doesn't the interest of PII lie in its consequences for the thesis that identity is primitive?

My reply is that there is more of interest in PII than its consequences for whether identity is primitive. PII states that there cannot be purely numerical difference, that is, that every numerical difference must be accompanied by an extra-numerical difference. This is a philosophically interesting thesis, even if an extra-numerical difference need not be a purely qualitative difference.

Some might think that, however interesting the thesis I am discussing is, it is not a version of the Identity of Indiscernibles, since 'the original motivation for [the Identity of Indiscernibles] involves the grounding of individuality in qualitative features of the world' (Ladyman and Bigaj 2010: 134; cf. Shumener 2021: 1031). But whose motivation is the original motivation for the Identity of Indiscernibles? The Stoics? Nicholas of Cusa? Leibniz? Someone else? And how much of the content of a thesis is fixed by its original motivation?

In any case, whatever the original motivation, the Identity of Indiscernibles is not a thesis about grounds—indeed, the Identity of Indiscernibles does not make any priority claim. The Identity of Indiscernibles is a supervenience thesis according to which there can be no numerical difference without some extra-numerical difference. But, as is well known, supervenience is too weak to give grounding. Thus, if the original motivation for the Identity of Indiscernibles was to ground individuality or identity in qualitative features of the world (a claim that still needs to be historically substantiated), that motivation was misguided, since the Identity of Indiscernibles, not making any priority claims, cannot provide such grounding.

That the Identity of Indiscernibles makes no priority claims but is simply a supervenience thesis should be obvious from its different formulations. The canonical Leibnizian formulation is 'There can be no difference *solo numero*', which makes no priority claim. Similarly, Leibniz often states the Identity of Indiscernibles in terms of resemblance: 'No two objects can resemble each other perfectly'—again, there is no priority claim here. It *might* be argued that for Leibniz numerical identity is *grounded* in the purely qualitative, but my point here is simply that his *formulations* of PII do not demand grounding. Also, the much discussed PIIa and PIIb, to be introduced in the next section, do not make any priority claims either. Finally, principle (1), which, as I have argued, is not a version of the Identity of Indiscernibles, but which is taken to be a trivial version of it by everyone else, does not make any priority claim either. All these formulations of the Identity of Indiscernibles claim simply that numerical difference must necessarily be accompanied by another kind of difference. Therefore, it should be no objection to PII, as I have formulated it above, that it doesn't ground identity or individuality in the qualitative features of the world.

But if PII does not require numerical identity to be grounded in the purely qualitative, does it not follow that PII cannot be the principle of individuation? And isn't the point of PII to provide a principle of individuation? Again, I think there is more of interest in PII than the fact that it might provide a principle of individuation. PII states that sharing all NT-properties is a sufficient condition for numerical identity. This is an interesting philosophical thesis in itself, whether or not it provides a principle of individuation. But whether it does, depends on what a principle of individuation is supposed to be and, as we shall see in the next section, there is at least one sense in which PII can provide a principle of individuation.

2.8 I shall now introduce two commonly discussed versions of PII. One of them restricts the domain of property quantification to pure properties, and the other one restricts it to intrinsic pure properties. Here they are, listed alongside PII:

> PII: Necessarily, no two objects share all their NT-properties.
>
> PIIa: Necessarily, no two objects share all their pure properties.
>
> PIIb: Necessarily, no two objects share all their intrinsic pure properties.

It is clear that PIIb is the strongest of the three principles, since it entails PIIa and PII, and neither of the latter entails it. And it is equally clear that while PIIa entails PII, the latter does not entail the former. Thus PII is the weakest of the three principles and it is indeed the weakest of all versions of the Identity of Indiscernibles, since it quantifies over all properties differing with respect to which is incompatible with purely numerical difference. Note that, since numerical difference is a symmetric but irreflexive relation, PII is not equivalent to the principle that demands that every two objects must be weakly discernible—by not imposing any restriction in the class of weakly discerning relations on which it quantifies, this latter principle becomes susceptible of trivial proof. Nevertheless, as I shall explain in Section 5.6, PII is equivalent to the principle that necessarily, every two objects are non-trivially weakly discernible.

Do any of the three versions of PII distinguished above provide a principle of individuation? There are different senses of the expression 'principle of individuation'. In one sense it is what makes an individual object an individual object. Clearly no version of PII provides a principle of individuation in this sense.

In another sense, a principle of individuation is what makes an individual object the individual object it is. But none of the principles distinguished above provides a principle of individuation in this sense. That it is necessary that no two objects share all of a certain set of properties does not guarantee that whatever properties from that set any object has, can only be had by that object. If some other object could have had the same properties a certain object has, then having those properties is not what makes the latter object the object it is.

What could provide a principle of individuation in this sense are the *trans-world* versions of the principles above. Such versions say that no objects, whether they exist in the same or in different possible worlds, can share all their NT-properties, all their pure properties, and all their intrinsic pure properties. Such trans-world versions are committed to the existence of *possibilia* but, as I pointed out in Section 1.6, in this work I am committed to an ontology that includes only actual objects. But, in any case, such versions are not PII, PIIa, and PIIb. Such versions entail, respectively, PII, PIIa, and PIIb, but PII, PIIa, and PIIb do not entail the respective trans-world versions, and so PII, PIIa, and PIIb do not provide a principle of individuation in the relevant sense even if the trans-world versions do.

There is a third sense in which a principle of individuation is what makes an individual object numerically different from every other one. Here the principles above fare differently.[9] Suppose PIIa and PIIb are true. Then the totality of the pure properties of an object, or intrinsic pure properties of an object, can be taken to be what makes the object numerically different from every other object. For, if PIIa and PIIb are true, no other object has the same pure properties, or intrinsic pure properties, any given object has.

But if PIIa and PIIb are false and PII is true, can PII provide a principle of individuation in the third sense? That is, if PII is true because, necessarily, no objects share all their *impure* NT-properties, can the non-shared NT-properties of an object be taken to be what makes that object numerically different from every other object? It might be thought they cannot be so taken because impure properties are those which consist in bearing a relation to a certain object in particular and therefore they presuppose that the *relata* are numerically different. But this is the case only when the impure property is based on an *irreflexive* relation. And the truth of PII need not be based in the fact that objects must differ with respect to at least an impure NT-property based on an irreflexive relation.

---

[9] The different senses of principles of individuation distinguished in this section are based on the discussion in King 2000. For a discussion of Leibniz's PII as a principle of individuation see Rodriguez-Pereyra 2014: 66–9.

Now, it is important to note two things. One is that claiming that PII provides a principle of individuation in this sense does not entail that the principle of individuation in question is the only principle of individuation in such a sense—that might or might not be the case, but it is an independent issue.

The other is that all I have argued is that a principle of individuation in the third sense distinguished above can be extracted from, or can be provided by, PII (and similarly for PIIa and PIIb)—in particular, I have not argued that PII is a principle of individuation in the third sense (and similarly for PIIa and PIIb). For principles of individuation make priority claims: they claim that something, whatever it is, is what grounds the individuation of certain objects (in whatever sense of individuation). For instance, in the third sense distinguished above a principle of individuation is what *makes* an individual object numerically different from every other one. But PII and its versions make no priority claim: all they claim is that there must obtain, between any two numerically different objects, a certain kind of extra-numerical difference. But they do not claim that numerical difference is grounded in the relevant kind of extra-numerical difference. Thus, although PII can provide a principle of individuation, it is not itself a principle of individuation.

3
Black's World

3.1 As I have said, my intention is to establish PII, and so I shall give two arguments for PII in Chapter 5. But if PIIa or PIIb are true, trying to establish PII is superfluous, since both PIIa and PIIb entail PII. But PIIa and PIIb are false, and so there is a point in trying to establish PII independently of PIIa and PIIb. The main argument against PIIa (and thereby it is an argument against PIIb too) is that the world could have contained just two objects sharing all their pure properties, intrinsic and extrinsic. The most prominent example of such a world appears in Max Black's famous paper on the subject, where Black considers the possibility that the world is such that the only objects are two iron spheres (and their parts) that have the same diameter, shape, temperature, colour, mass, exact chemical composition, and so on (Black 1952: 156). Now, although Black does not use the terminology of pure and impure properties, he effectively characterizes what I shall call *Black's world* as a situation in which the spheres share all their pure properties: 'every quality and relational characteristic of the one would be a property of the other' are Black's words (1952: 156), and he clearly means pure qualities and pure relational characteristics. Let us call the possibility in which the only existing objects are two iron spheres that share all their pure properties, *Black's world*.

If Black's world is possible, then PIIa, and consequently PIIb, is false. I shall argue that Black's world is possible in the next chapter. But whether any possibility in which the only existing objects are two iron spheres is a possibility in which those spheres share all their pure properties can, and has, been questioned. Indeed, some have argued, or can be construed as having argued, that no possibility in which the only existing objects are two iron spheres is one in which those spheres share all their pure properties. Thus, in this chapter I shall argue against actual or possible arguments that Black's world is impossible because if there were only two iron spheres, they would have to differ with respect to some pure properties. Showing that such arguments do not work does not amount to showing that Black's world is possible—showing that Black's world is possible is, as I said, the task of the next chapter.

3.2 In Black's world the only concrete objects are two iron spheres (and their parts) having the same diameter, shape, temperature, colour, mass, exact chemical composition, and so on. It might be thought, however, that they must differ with respect to the pure relational properties that derive from relations to abstract objects. But the presence of abstract objects in the situation where the spheres are the only concrete objects does not make them differ with respect to any pure properties. I guess this is why, despite the fact that, presumably, abstract objects—or at least some of them—are necessary entities that cannot fail to exist, Black does not explicitly make the restriction to concrete objects when specifying his world. In any case, let us see why abstract objects do not make the spheres differ with respect to any pure properties.

For abstract objects to make the spheres differ with respect to pure properties, it would have to be the case that one of the spheres, say sphere a, was related in a certain way to an abstract object c having certain pure properties, while the other sphere, sphere b, was not related in that way to any abstract object having the same pure properties as c. But this is not the case with Black's spheres. Thus if one of the spheres in Black's world is related in a certain way to an abstract object c that has certain pure properties, so is the other sphere also related to an abstract object that has the same pure properties as c, whether this object is c or not (this generalizes to parts of the spheres having the same pure properties). Therefore, the presence of abstract objects in Black's world does not entail a difference in the pure properties of the spheres.

Let us confirm this point. There is no variation between the spheres in how they relate to mathematical or geometrical objects unless they differ with respect to shape, size, temperature, mass, and so on. Thus, the spheres are equally related to every mathematical and geometrical object. For instance, assuming that 1 is the number of miles in the diameter of one of the spheres, 1 is also the number of miles in the diameter of the other sphere, and therefore both spheres are related in this way to number 1. And, I say, they are equally related to number 1 in every other way they are related to number 1. And, in general, they are equally related to every mathematical and geometrical object.

Similarly, since they have the same pure properties, the spheres are equally related to every pure property. But, clearly, they are not equally related to all impure properties. For instance, only sphere *a* has the property of *being identical with a*, and so only sphere *a* bears the relation of *having* to the property of *being identical with a*. But the property of *being identical with a* shares its *pure* properties (for instance, the property of *being a property of identity*) with the property of *being identical with b*, the property of identity to which sphere *b* is uniquely related through the relation of *having*. Thus, the difference of the spheres with respect to such impure properties does not make them differ with respect to their pure properties (for example, both of them share the property of *having a property of identity*). This generalizes to all other impure properties the spheres are not equally related to.

A similar point can be made with respect to sets. The spheres are equally related to all pure sets. But they are unequally related to impure sets that contain only one of them as members and even to some impure sets that contain neither of them as members but contain one or both of them in their transitive closure, for instance sets like $\{\{a\}\}$ and $\{\{a\},\{a,b\}\}$. Consider first sets that have only one of them as member—for instance, only sphere *a* is a member of $\{a\}$ and only sphere *b* is a member of $\{b\}$. But, as I argued in Section 2.5, *being a member of $\{a\}$* and *being a member of $\{b\}$* are impure properties. And by being differently related to those sets, the spheres *a* and *b* do not acquire different pure properties, since $\{a\}$ and $\{b\}$ have the same pure properties, for instance they share the property of *being a singleton set whose member is an iron sphere located two miles away from another iron sphere*, and so on. Thus both *a* and *b* have the pure property of *being a member of a singleton set whose member is an iron sphere located two miles away from another iron sphere*, and so on. Obviously this point generalizes to all other sets that contain only one of them as members. It should now also be clear that being differently related to sets that contain neither of them as members but contain one or both of them in their transitive closure does not make the spheres have different pure properties. For the way in which a is related to {{a}} and {{a},{a,b}} is the way in which b is related to {{b}} and {{b},{a,b}}. And {{a}} and {{b}} have the same pure properties and so do {{a},{a,b}} and {{b},{a,b}}. And this generalizes to all other sets that contain one or both of them in their transitive closure. The point is a general one: for every set containing a at some point in its transitive closure, there is another set that differs from it only by having b at that point in its transitive closure, and those sets will have exactly the same pure properties.

Similarly, if space in Black's world is absolute, the spheres are not equally related to the spatiotemporal regions occupied by them: only sphere a occupies the region occupied by sphere a and only sphere b occupies the region occupied by sphere b. But *occupying region R* and *occupying region R** are impure properties. And by being differently related to those regions, the spheres a and b do not acquire different pure properties, since regions R and R* have the same pure properties, for instance they share the property of *being two miles away from a region occupied by an iron sphere*, and other such properties. Thus both a and b have the pure property of *occupying a region that is two miles away from a region occupied by an iron sphere*, and other such properties. Thus, the difference of the spheres with respect to those regions does not make them differ with respect to their pure properties. Similarly, the difference of the spheres with respect to any spatiotemporal points does not make them differ with respect to their pure properties.

(Needless to say, the same is true if space is not an object, but having spatial relations is a matter of bearing spatial relations to other concrete objects. In that case the spheres would still differ with respect to some of their spatial impure properties, for instance the property of *being two miles away from sphere b*. But they would nevertheless have the same spatial pure properties, properties like *being two miles away from a sphere having such and such a temperature* or *being at no distance from a sphere having such and such a mass*.)

Similarly, the spheres cannot be distinguished in the relevant sense by appeal to vectorial differences. Each one of them will have, for instance, pure properties like *being at the terminal point of a vector at whose initial point lies a sphere of such and such a mass*. What they will differ with respect to is impure properties like *being at the terminal point of a vector at whose initial point lies sphere b*.

The same point can be made with respect to bare particulars, assuming that the spheres have bare particulars and such bare particulars are abstract objects. In that case each sphere is uniquely related to its own bare particular. But, again, this does not make a difference with respect to the pure properties of the spheres, since both bare particulars have exactly the same pure properties. For example, both bare particulars share the pure properties of *being an abstract object, being a bare particular, being a bare particular of an iron sphere of such and such diameter*, and similar ones.[1]

We have seen that whenever one of the spheres is related in a certain way to a mathematical or geometrical object, a property, a set, a spatiotemporal region or point, a vector, or a bare particular, either the other sphere is related in the same way to the same object, or it is related in the same way to another object having the same pure properties as the object to which the first sphere is related. And it is plausible that the abstract objects I have considered exhaust the types of abstract objects. And even if there are abstract objects I have not considered, it is plausible that the same considerations apply to them as they apply to the abstract objects I have considered. Indeed, given the exact similarity of the spheres and the symmetry of Black's world, how could one of the spheres be related to any abstract object with certain pure properties without the other sphere being related to an abstract object with the same pure properties? Thus, the presence of abstract objects in Black's world does not make a difference in the pure properties of the spheres. Nor does it make a difference in the pure properties of the parts of the spheres, since to every part of one of the spheres there is a corresponding part in the other sphere with the same pure properties, and the points I have made about the relations

---

[1] Thus, bare particulars have properties, despite the common assertion that they don't. That they have properties on my conception of properties as predicable conditions should be obvious. But even on a conception of properties as concrete universals they should be conceived as having properties. Indeed, discussing bare particulars in the context of such a theory of universals, Sider argues that bare particulars may be red, round, and juicy (Sider 2006: 388). Of course, having such universals would make bare particulars concrete objects, while I am here assuming, for the sake of argument, that they are abstract objects. Despite this difference, the point stands: bare particulars have properties.

between the spheres and abstract objects equally apply to the relations between the parts of the spheres and abstract objects.

Therefore, the presence of abstract objects in Black's world does not make a relevant difference to the truth or falsity of PIIa and PIIb, and so from now on I shall speak of Black's world as containing just two iron spheres—without bothering to make the restriction to concrete objects explicit. Also, although Black's world is supposed to contain the spheres *and their parts*, I shall often drop the reference to their parts, and the reader should take that as implicit—I shall, however, return to the issue of the parts of the spheres in Section 4.5, where their discussion is pertinent.[2]

3.3 Could necessary concrete objects make a difference in the pure properties of the spheres? My description of Black's world assumes that if it is possible, there are no necessary concrete objects (or at least no necessary concrete objects other than the spheres!). Although I do not believe in any necessary concrete objects, I cannot disprove their existence either. But the assumption is innocuous, since there is only one alleged necessary concrete object that is even remotely plausible to believe in, namely God, and there is no reason to believe that God would be asymmetrically related to the two iron spheres if the spheres were the only other concrete objects and they shared all their pure properties. It is absurd to believe that God, for instance, would love one of the spheres but not the other. Indeed, it is absurd to believe that God would be intentionally related in a certain way to only one of the spheres.

---

[2] Must the spheres have parts? A world like Black's but in which the spheres are extended simples seems to be a possible one. In fact, I have argued that one of Leibniz's arguments for the Identity of Indiscernibles fails precisely because the argument is unable to rule out a world like Black's containing two simple indiscernible objects, whether or not they are *extended* indiscernible simples (Rodriguez-Pereyra 2014: 108–16). Now, since extended simples are more controversial than extended complexes, the difference in dialectical situation is a relevant difference. If one is trying to refute an argument for a version of the Identity of Indiscernibles, which is what I was trying to do in the book on Leibniz I just cited, it is sufficient to appeal to a world of simples, leaving open whether such simples are extended; but if one is trying to establish the falsity of a version of the Identity of Indiscernibles, which is what I shall do in the next chapter, one should aspire to more than merely assuming the possibility of extended simples and then establish the possibility of a world with two indiscernible extended simples. Thus, I shall establish the possibility of a world with two indiscernible *complex* spheres, spheres having proper parts.

Now, some might disagree and not consider absurd the supposition that God might be differently intentionally related to the spheres. Never mind. A world in which God is not differently related to the spheres is certainly possible, and in such a world the presence of God does not make a difference to the pure properties of the spheres. Therefore, even if God is a necessary concrete object, it does not follow that any two iron spheres must differ with respect to their pure properties.

Now, Leibniz argued that, if there were indiscernible objects, God would have been asymmetrically intentionally related to them by having decided to put them in different places (Leibniz 1956: 61; see also Rodriguez-Pereyra 2014: 104–17). But it should be clear from the discussion in the previous section that this would not create a difference in the *pure* properties of the spheres. Nevertheless, even if God could be asymmetrically related to the spheres by loving one and not the other, which would create a difference in the pure properties of the spheres, this need not be the case, as I pointed out above. And so, even if God is a necessary concrete object, there is a world in which God is related to the spheres in such a way that it does not make a difference in the pure properties of the spheres. Now, such a world is not Black's world, since I defined Black's world as a world in which the only concrete objects are two iron spheres, and in this one there is a third object, namely God. But never mind: since in this world God's relations to the spheres does not produce a difference in the spheres' pure properties, God's presence does not make a relevant difference to the truth or falsity of PIIa and PIIb. Therefore, for simplicity, I shall from now on continue to think of Black's world as one in which the only concrete objects are the two iron spheres (I shall briefly come back to the issue of a necessary God in Section 4.5).

3.4 The spheres were stipulated to have exactly the same physical and chemical pure properties like their diameter, shape, temperature, colour, mass, exact chemical composition, and so on. We have seen that neither abstract objects, nor God, nor space or time or spacetime, make a difference in the pure properties of the spheres. And we also saw that they have the same pure properties that derive from their spatial relations to each other. Now, in Black's world the spheres coincide exactly temporally: they exist for exactly the same period of time, whatever that period is. They also bear no causal relations to each other; that is, neither sphere is formed from matter of the other, nor did one collide into the other making the latter move in a certain way, nor have they ever come into causal contact in any way. That means that the spheres do not differ with respect to pure properties that derive from spatiotemporal and causal relations.[3]

Must then they differ with respect to some of their modal pure properties? It is clear that being objects of the same kind, they share all the modal pure properties that specify what is metaphysically necessary or possible for them, that is, properties of the form *being necessarily F* or *being possibly F*, where F is a predicate standing for a pure property. But these are not the only types of modal properties, and some have argued that the spheres *must* differ with respect to some modal properties. I shall now consider some of those arguments.

It is metaphysically possible that there is a particle near one of the spheres but not near the other. Thus, given their mass, the spheres have different dispositions to affect a particle that, if it existed, would be near the one but not near the other. This led Ronald Hoy to argue that only one of the spheres has the disposition to affect that particle, if it were near one of the spheres but not near the other, in a certain way (Hoy 1984: 291).

Now, if so, with respect to which property do the spheres differ? It might be thought that they would differ with respect to the property of *being disposed to affect in such a way any particle that is near it but not near the other sphere*. But this is not the case: both spheres have that property. For sphere *a* has the impure property of *being disposed to affect in such a way any particle that is near it but not near sphere b*, and sphere *b* has the impure property of *being disposed to affect in such a way any particle that is near it but not near sphere a*. Thus both spheres have the

---

[3] That the spheres in Black's world do not bear any causal relations to each other is not essential to the philosophical significance of Black's world, namely refuting PIIa. The spheres could be causally related to each other in several ways, provided they were symmetrically causally related and neither was causally related to itself in a way the other one was not causally related to itself—in that case no difference in pure properties would arise from these causal relations. But note that if the spheres are causally related, causation must be a primitive relation, not accountable in terms of differences in pure properties or spatiotemporal relations. For a discussion of primitive causation relevant to the present topic, see Audi 2011.

pure property of *being disposed to affect in such a way any particle that is near it but not near the other sphere*.

Hoy (1984: 291) is aware of the symmetry of the spheres with respect to that property. But he suggests that since there could have been a certain individual particle near one of the spheres but not near the other, the spheres differ in the property of *being disposed to affect in such a way that particle*. Putting aside the fact that I am assuming that Actualism is necessarily true, according to which supposition it is not even possible that there are merely possible objects, there are two problems with Hoy's suggestion. One is that the property of *being disposed to affect in such a way that particle* is impure, since it consists in bearing a disposition to a certain object in particular. The other problem is that both spheres would have that property, since presumably that individual particle could also have been near the other sphere.

Similarly, if space is an object, the spheres differ with respect to the property of *being disposed to affect any particle that occupies region of space R*, where R is a region that is near one of the spheres but not the other. But, as is clear, such property is impure.

Jeremy Goodman has recently provided an argument that, like Hoy's, tries to distinguish the spheres of Black's world by means of modal properties. Goodman argues, on the basis of Conditional Excluded Middle, that the spheres in Black's world are discernible with respect to their pure properties because only one of them has the property of *being a duplicate of an iron sphere such that, had one of these two spheres been heavier than the other, it would have been the heavier one*.[4] Conditional Excluded Middle is the principle that either, if it were the case that p, it would be the case that q or, if it were the case that p, it would be the case that *not-q*. One instance of Conditional Excluded Middle is that either,

---

[4] Three clarificatory points. First, although Goodman speaks of qualitative properties rather than pure properties, his qualitative properties are my pure properties (see Goodman 2015: 1, fn. 1). Second, Goodman assumes that the spheres in Black's world are orbiting each other, and so the property he considers is *being a duplicate of an iron sphere that it is orbiting such that, had one of these two spheres been heavier than the other, it would have been the heavier one*. But Black never stipulated that the spheres were orbiting each other (Black 1952: 156). So, to be faithful to Black, I shall represent Goodman's point using the property I mentioned in the main text. Third, although Goodman does not define 'duplicate', duplicates are often taken to be (but not necessarily defined as) objects that do not differ with respect to their intrinsic pure properties, and this is how I shall understand the word.

if it were the case that one of the spheres was heavier than the other, sphere *a* would be the heavier one or, if it were the case that one of the spheres was heavier than the other, it would be the case that sphere *a* is not the heavier one. With the help of a few other uncontroversial premises (see Goodman 2015: 3), Goodman infers that the spheres must differ with respect to the counterfactual property of *being a duplicate of an iron sphere such that, had one of these two spheres been heavier than the other, it would have been the heavier one*.

Now, Conditional Excluded Middle is a controversial principle—it has been rejected by Lewis (1973), among others—and one way of resisting the argument is by rejecting Conditional Excluded Middle. But, although available, this route is not necessary, since there are other ways of resisting the argument. For instance, if the spheres had all their intrinsic pure properties necessarily, they could not have differed in mass, and so it would not follow that the spheres differ with respect to the counterfactual property. Thus, one need only appeal to a world like Black's containing spheres that have all their intrinsic pure properties necessarily. But this is not necessary either. One might simply appeal to a world like Black's containing spheres such that, necessarily, if they coexist, they must have the same intrinsic pure properties. Again, if this were the case, the spheres could not have differed in mass, and so it would not follow that the spheres differ with respect to the counterfactual property.

If one has doubts that it is possible for any spheres to have all their intrinsic pure properties necessarily or that it is possible for any spheres to be such that, necessarily, if they coexist, they must have the same intrinsic pure properties, there is still another way of resisting the argument. For given that the spheres are indiscernible with respect to their intrinsic pure properties, and in particular their mass, the right thing to say is that it is indeterminate of each sphere whether it has or lacks the counterfactual property mentioned above. If so, the spheres do not determinately differ with respect to that property.

Goodman considers a similar response, but his reply is that all it shows is that one should distinguish between, on the one hand, the principle that, necessarily, every two objects differ with respect to their pure properties and, on the other hand, the principle that, necessarily, every two objects *determinately* differ with respect to their pure properties. He associates the former with the traditional Identity of Indiscernibles and so the fact that his argument does not affect the latter does not bother him (Goodman 2015: 7).

But this is incorrect. It is true that the traditional formulations of PIIa do not mention determinacy. But everyone who has thought about PIIa has thought that it requires that numerically different objects must *determinately* differ with respect to their pure properties. This is what Black's world has been thought to refute. Adding a determinacy operator is not bringing in another principle, but making explicit what was implicit in the original principle.[5] And the reason why Black's world has been thought to refute the original principle is precisely that in Black's world the spheres do not *determinately* differ with respect to their pure properties. Thus Goodman's argument does not show that Black's spheres must differ in the sense in which they were supposed not to differ.

Goodman also thinks that Conditional Excluded Middle supports what he calls the necessity of discernibility, the thesis that, necessarily, objects that are indiscernible with respect to their pure properties are necessarily indiscernible with respect to their pure properties (Goodman 2015: 16–17). And Goodman says that he does not have a proof of the Principle of Identity of Indiscernibles (by which he means PIIa) because he does not have a proof that there are no objects that are necessarily indiscernible (that is, necessarily indiscernible with respect to their pure properties). But he says that it is hard to imagine what such pairs of objects would be like, and that such extreme necessary connections between distinct existences is hard to fathom (Goodman 2015: 4).

But that the spheres share all their pure properties, including their modal ones, does not entail that they are necessarily indiscernible with respect to their pure properties. The property of *possibly being the only existing object* is a pure modal property, and both spheres determinately have it. Similarly, they both determinately have the properties of *possibly being the only iron sphere, possibly being the only sphere with such and*

---

[5] Because the idea of determinacy was implicit I cannot provide any quotes of authors stating the principle with a determinacy operator or an explicit reference to determinacy. But one should expect the reference to determinacy to be implicit since it is only against the background of a discussion questioning determinacy that philosophers make explicit reference to determinacy.

such pure properties, and so on. But if they determinately have these properties, they are not necessarily indiscernible with respect to their pure properties. For instance, if both spheres have the property of *possibly being the only existing object*, it is possible for each to be the only existing object, but then they are not necessarily indiscernible, since they would not share the pure property of *existing* if only one of them existed. Similarly, if the spheres have the property of *possibly being the only iron sphere*, it is possible for only one of them to be an iron sphere, and therefore they are not necessarily indiscernible, since they would not share the pure property of *being an iron sphere* if only one of them were an iron sphere.

As we shall see in Chapter 5, the thesis that indiscernible objects are necessarily indiscernible is true when indiscernibility is not restricted to pure properties but it embraces all NT-properties, including impure ones.

3.5 We have seen that the spheres of Black's world do not differ with respect to any of their pure properties, whether intrinsic or extrinsic, whether modal or not, and that they only differ with respect to their impure properties. But mustn't they differ with respect to pure relations? And, if so, isn't this a purely qualitative difference?

That is what Fred Muller believes. Indeed, he has argued that, although Black's spheres are not absolutely discernible, they are relationally discernible—that is, he has argued that, although the spheres do not differ with respect to any properties, they differ relationally (Muller 2015: 210; cf. Robinson 2000: 170). Differing relationally, for Muller, consists in there being a pure relation that discerns the objects in question (Muller 2015: 206). Thus, more precisely, Muller's idea is that the spheres are weakly discernible, since they differ with respect to a symmetric but irreflexive relation, the relation of *being two miles away from*.[6] There is no reason not to let the lambda-operator abstract relations as well as properties, in which case we could express the relation of *being two miles away from* as follows: $(\lambda xy)(x$ is two miles away from $y)$. This is a pure relation because entering into it does not consist in being related to any object in particular. Are there impure relations? Yes. Here is one

---

[6] Interestingly, Adrian Heathcote (2022: 370) has pointed out that Black was aware of the possibility of this point.

example: $(\lambda xy)(x$ is two miles away from y and the Eiffel Tower is equidistant from x and y). The fact that for two objects to stand in this relation is for them not only to be two miles away from each other but also to be equidistant from a certain tower in particular makes this relation impure.

That Muller's concern is with *pure* properties and relations should be clear from his response to the claim that the spheres differ with respect to the properties of *being two miles away from Castor* and *being two miles away from Pollux*.[7] His response is that it is not permissible to discern the spheres through such properties because such properties exploit the fact that the spheres bear names (Muller 2015: 215). Indeed, he is explicit that predicates in which names occur should not be permitted to discern presumably indiscernible objects (Muller 2015: 227), which indicates that his concern is with pure properties and relations.

If Muller is right, the spheres in Black's world differ purely qualitatively without differing with respect to any pure property. If so, even if the spheres share all their pure properties, they differ purely qualitatively.

But to bear a relation, whether pure or impure, is to bear it to some object or objects—nothing can bear a relation without bearing it to any objects. Therefore, for two objects to differ with respect to any relation, whether a pure or impure one, is for them to differ with respect to *which* object or objects they are related to by that relation (this is the case both if they differ with respect to a relation because they are related by it to different objects and if they differ with respect to a relation because only one of them is related by it to some object or objects). Therefore, if two objects differ with respect to a relation, they do not thereby differ purely qualitatively, *even if the relation in question is pure*. Thus, even if the relation of *being two miles away from* is a pure relation because entering into that relation with some object does not consist in being related to any object in particular, for two objects to differ with respect to it is for them to differ with respect to which object in particular they are related to by that relation. Indeed, for the spheres to differ with respect to the relation

---

[7] 'Castor' and 'Pollux' are the names that the spheres receive briefly in Black's dialogue (Black 1952: 158). Except when discussing philosophers who refer to them as Castor and Pollux, I shall always refer to the spheres as a and b.

of *being two miles away from* is for one of them to bear it to Castor and not to Pollux and for the other one to bear it to Pollux and not to Castor.

Thus the relational difference between the spheres can be captured in terms of impure properties: the spheres differ with respect to the impure properties of *being two miles away from Castor* and *being two miles away from Pollux*. The relational difference between the spheres can also be captured in terms of relational facts. For instance, it is a fact that Castor is two miles away from Pollux but it is not a fact that Castor is two miles away from Castor. But if we extend the pure/impure distinction from properties to facts, such facts are impure. Indeed, Muller himself represents the difference between Castor and Pollux by saying that the following obtain: D(C,P), D(P,C), not-D(C,C), and not-D(P,P), where 'D' stands for the relation of *being two miles away from* and 'C' and 'P' stand for Castor and Pollux respectively (Muller 2015: 210). But here the particular objects Castor and Pollux enter into the relevant facts that discern them. Muller (2015: 227) says that one is not allowed to discern the spheres in terms of predicates where names occur (or, rather, in terms of impure properties), but then why does he discern the spheres in terms of sentences where names occur (or, rather, in terms of impure facts)? The truth is that the relational difference between the spheres is not a purely qualitative one, but an impurely qualitative one.

Thus, although the spheres differ with respect to a pure relation, this does not make them differ purely qualitatively. Two consequences follow from this. First, since the spheres of Black's world share all their pure properties, as I argued earlier in this chapter, they are entirely purely qualitatively identical. Therefore, if Black's world is possible, objects can be entirely purely qualitatively identical. Second, since no difference with respect to a relation produces a purely qualitative difference, there is no need to quantify over pure relations to have a version of PII that rules out purely qualitatively identical objects—PIIa is fine to that effect.

4
The Possibility of Black's World

4.1 Since the spheres in Black's world share all their pure properties, intrinsic and extrinsic, modal and non-modal, if Black's world is possible, PIIa and PIIb are false. But is Black's world possible? I shall argue in this chapter that it is indeed possible.

Why do I need to argue that Black's world is possible? Didn't Black argue for that? In Black's paper (1952: 156) there is no more than an appeal to our imaginative intuition that it is possible that the universe should have contained nothing but two exactly similar iron spheres, and it is clear that Black has in mind exact similarity with respect to pure properties. Black, or, more precisely, one of the participants in his fictional dialogue, says that such a situation is *logically* possible, but the argument can be taken as intending to establish the metaphysical possibility of objects indiscernible with respect to their pure properties; this is how it is usually understood in the contemporary discussion, and this is how I am going to understand it. Thus, I shall, as is commonly the case, interpret the example in Black's paper as one of metaphysical possibility and as intended to refute PIIa.

Now, Black's thought experiment has been criticized because, it is argued, the imagined situation can be redescribed as containing only one sphere, rather than two indiscernible ones. Indeed, that there is a possible world where there is an iron sphere having such and such a diameter, shape, temperature, colour, etc., and which is at a certain distance from an iron sphere having the same diameter, shape, temperature, colour, etc., leaves underdetermined whether in that world there are two indiscernible spheres or one sphere that is at a distance from itself, that is, a bi-located sphere. Thus, it is concluded that it remains open how to correctly describe the content of our imagination (Hacking

1975: 251; cf. O'Leary-Hawthorne 1995: 195).[1] It is important to note that the objection is not to the link between imaginative conceivability and possibility, a link presupposed by Black's argument. The objection is, rather, to the presupposition that the content of the act of imaginative conception is a situation with two spheres. The objection is pressing, for if the imagined scenario can be described as containing just one sphere, then it is not clear that one is conceiving a situation with two spheres, and therefore it is not clear that there is a possible world containing just two spheres that share all their pure properties. As Ian Hacking says, in arguing that in a certain possible world there exist two distinct but indiscernible objects, bland assertion that there are two such objects is not enough; there must be argument (Hacking 1975: 251).

I used to think that a version of Adams' famous and influential argument from almost indiscernibles (Adams 1979) could be used to establish the possibility of Black's world. The thought was that one can imaginatively conceive a possible world containing two almost indiscernible iron spheres, say, a possible world exactly like Black's except that the spheres in it differ infinitesimally in temperature—there is no question that what one is imagining here is a situation containing two spheres, since one is imagining a sphere with a certain temperature and a sphere with a different temperature. But if that world is possible, then a world where those two spheres have the same temperature and everything else is the same as in the world where they differ infinitesimally, is also a possible world. This last world is Black's world and it cannot be correctly described as a world containing just one sphere at a distance from itself (Rodriguez-Pereyra 2004: 74).

But I subsequently came to reject Adams' argument (Rodriguez-Pereyra 2017a). Furthermore, as Martin Pickup has pointed out to me, since it is possible that a sphere is in two places, it is possible that a sphere has two different temperatures, and so this argument is also vulnerable to a version of Hacking's objection. If Adams' argument from

---

[1] There is a remark in van Cleve (1985: 104) that aligns well with the spirit of Hacking's objection, without amounting to a version of the objection. For discussion of Hacking's objection, or Hacking-style objections to Black's thought-experiment, see, among others, Adams 1979, French 1995, Valicella 1997, Zimmerman 1997, Hughes 1999, Rodriguez-Pereyra 2004, Cross 2011, and Curtis 2014.

almost indiscernibles does not work, is there any reason to think that Black's world is possible?

In what follows I shall provide an argument that Black's world is metaphysically possible. The argument will not appeal to an imaginative conception of the scenario realized by Black's world, and therefore it will not be vulnerable to Hacking-like objections.

But is this really necessary? Recently, Calosi and Varzi (2016) have argued that defending PII from Black's argument by arguing that the alleged world with two indiscernible spheres contains a bi-located sphere adds nothing to the simple denial that there can be indiscernible objects. This is because defending PII in such a way, they argue, forces one to identify a world with a single, bi-located sphere, with a world with a single, uniquely located sphere (Calosi and Varzi 2016: 10). Nevertheless, I still think that the enterprise I shall take in what follows is necessary. For even if Calosi and Varzi are right that appealing to bi-location adds nothing to the simple denial that there can be indiscernible objects, rejection of PIIa on the basis of Black's world still requires avoiding Hacking-like objections that the world one has described might contain only one sphere.[2]

I shall proceed in two stages. First, I shall give an argument that PIIb is false, and then I shall give an argument, based on the conclusion that PIIb is false, that Black's world is possible, which entails that PIIa is false. This argument is not vulnerable to Hacking's objection. Before advancing my argument against PIIb, I shall discuss and reject a couple of other arguments that attempt to establish the falsity of PIIb.

---

[2] For the record, I am not convinced by Calosi and Varzi's argument. This is for two reasons. The first is that they assume that the defender of the Identity of Indiscernibles will adopt a conception of possible worlds which makes them fall under the range of Identity of Indiscernibles (Calosi and Varzi 2016: 4, 6, 9–10). But there is no reason why the defender of the Identity of Indiscernibles must make possible worlds fall under its range, since there is no reason why the defender of the Identity of Indiscernibles must reify possible worlds or conceive of them as concrete objects. The second reason is that, although they don't use the lambda-operator device to refer to properties, they implicitly assume (Calosi and Varzi 2016: 7) that if Castor is at a distance from itself, this must be interpreted as attributing to Castor the property $(\lambda x)(x$ is at a distance from Castor$)$; but it can also be interpreted as attributing to Castor the property $(\lambda x)(x$ is at a distance from $x)$—and interpreting such properties in this way allows a world with a single bi-located sphere to be distinguished from a world with a single uniquely located sphere.

4.2 Many have argued that, as a matter of fact, there are intrinsically indiscernible objects, since there are intrinsically indiscernible particles (see, among others, Cortes 1976, French 1989, and Erdrich 2020: 36). If it is indeed correct that there are objects sharing all their intrinsic pure properties, then PIIb is false, since what is actual is metaphysically possible. What I shall call the *First Argument from Science* can be put in this very simple form:

(1) Physical particles share all their intrinsic pure properties.
(2) What is actual is metaphysically possible.
(3) Therefore, PIIb is false.

The principle that what is actual is metaphysically possible is unobjectionable. Furthermore, particles are theoretical, unobservable entities, postulated because they play an explanatory role, and the properties attributed to them are exactly those needed to play the explanatory role. Attributing to particles no more properties than those needed to perform the explanatory role is good and standard methodology, since it is the application of Ockham's Razor to the case of properties: do not attribute properties to entities without necessity. The intrinsic pure properties attributed to particles by science do not distinguish between them, and if the intrinsic pure properties attributed to particles by science are all their intrinsic pure properties, then Premise (1) of the First Argument from Science is correct.

But there is a big gap between not attributing properties to particles without necessity and denying of particles any properties other than those it is scientifically necessary to attribute to them. And it is this gap that the supporters of PIIb will exploit to reject the argument. My point is not that assuming that the only intrinsic pure properties of particles are those attributed to them by science is not justified. My point is that the gap just identified offers the supporters of PIIb an opportunity to resist the argument.

There is another argument from science, suggested to me by Claudio Calosi, which I shall call the *Second Argument from Science*:

(1) It is physically possible that particles share all their intrinsic pure properties.

(2) What is physically possible is metaphysically possible.
(3) Therefore, PIIb is false.

The second premise of this argument is unobjectionable, and the first premise can be maintained on the basis that the laws of nature constrain the properties particles can have, but such laws are not incompatible with there being particles that share all their intrinsic pure properties. That the laws of nature constrain the properties of particles is important, since mere compatibility with the laws of nature should not guarantee physical possibility—otherwise whatever is not constrained by the laws of nature, that is, anything the laws 'don't speak about', would be automatically physically possible and therefore metaphysically possible too.[3] In any case, I think there is less margin for those endorsing PIIb to resist the Second Argument from Science.

Now, our knowledge of the laws of nature is scientific knowledge and there is no guarantee that science might not discover in the future that the laws of nature are incompatible with there being particles that share all their intrinsic pure properties. Similarly, there is no guarantee that science might not discover in the future intrinsic pure properties with respect to which all particles must differ. Thus, basing the rejection of PIIb on either of the arguments from science is too conditional on the current state of science. Those who endorse PIIb might see this as a reason to reject both arguments. I see it as a reason to search for a firmer argument against PIIb.

4.3 Consider Lewis' Recombination Principle, according to which putting together parts of different possible worlds yields another possible world (Lewis 1986: 87–8). For instance, to follow an example of Lewis', if there is a possible world where a dragon exists, and there is a possible world where a unicorn exists, there is a possible world where (a duplicate of) the dragon and (a duplicate of) the unicorn exist (Lewis 1986: 88). Now, since for Lewis duplicates have the same intrinsic (pure) properties (Lewis 1986: 62), given the Recombination Principle, PIIb is

---

[3] I am indebted to Claudio Calosi and Erica Shumener for this point.

false.[4] This suggests the following argument, which can be called the *Recombination Argument*:

(1) In world $w1$ there exists an iron sphere having intrinsic pure properties F, G, H, etc., and in world $w2$ there exists an iron sphere having intrinsic pure properties F, G, H, etc.
(2) There is a world $w3$ where a duplicate of the sphere in $w1$ exists and a duplicate of the sphere in $w2$ exists.
(3) Therefore, PIIb is false.[5]

Even granting the two premises and the Recombination Principle (by which the second premise follows from the first), the argument faces two serious problems.

First, note that in the context of Modal Realism, the context in which the argument is formulated, there is a version of the Principle of Identity of Indiscernibles that is stronger than PIIb. This is the trans-world version of PIIb, which I discussed in Section 2.8. This principle can be formulated as follows: if x exists in world w and y exists in world w^* then x and y do not share all their intrinsic pure properties, whether or not $w = w^*$. This principle thus rules out both trans-world and intra-world indiscernible objects. And in the context of Modal Realism, in which reality consists of the whole plurality of possible worlds, there is reason to focus on the stronger principle ruling out both trans-world and intra-world indiscernible objects (which entails PIIb, which rules out only intra-world indiscernible objects). Indeed, given that *possibilia* are fully real according to Modal Realism, there is a sense in which there being no intra-world indiscernible objects in any possible world does not rule out there being indiscernible objects *simpliciter*, since there might be trans-world indiscernible objects. Thus, in the context of Modal Realism, those who want to exclude intrinsic indiscernible objects from the whole

---

[4] Obviously, Lewis' intrinsic properties are all pure, since duplicates differ with respect to intrinsic impure properties, properties like *being identical with Aristotle* and *having Yorkshire as a part*.

[5] James van Cleve (2002: 391) noted the possibility of arguing in roughly this way against what he calls Strong PII, which is roughly my PIIb, but he neither endorsed nor rejected the argument.

of reality should adopt the stronger principle. But premise (1) assumes the falsity of the stronger principle, since the two spheres referred to in that premise are trans-world intrinsically indiscernible objects.

It might be thought that basically the same argument can be run without assuming the falsity of the stronger principle. The reason why this might be thought is that Lewis says that not only any two objects admit of combination but also 'any possible individual should admit of combination with itself: if there could be a dragon, then equally there could be two duplicate copies of that dragon side by side, or seventeen or infinitely many' (Lewis 1986: 89). Thus, one might think that one could just appeal to the existence of an iron sphere with certain intrinsic pure properties in the actual world, and pass from there to a world in which there are two duplicates of the sphere in question, thereby establishing the falsity of PIIb. Nevertheless, this version still assumes the falsity of the stronger principle, since the spheres in the second world are indiscernible from the sphere in the actual world.

The second problem is that since the spheres in worlds *w1* and *w2* are duplicates of each other, any duplicate of either is a duplicate of the other, and so there is no guarantee that in *w3* there are two spheres rather than one sphere at a distance from itself—in other words, the situation that obtains in *w3* can be described as one where there is just one sphere at a distance from itself, rather than two spheres. Thus, this argument is vulnerable to the same kind of criticism that Hacking advanced against Black's thought-experiment.

As I said in Chapter 1, I am committed to an actualist point of view rather than a modal realist one. Having seen how and why the Recombination Argument fails in the context of Modal Realism, one might wonder whether it can be formulated in an actualist setting. Of course the actualist does not have merely possible spheres in different possible worlds to combine with each other in a third different possible world, for there are no merely possible spheres in the actualist ontology, but one might think that the actualist might make use of Lewis' idea that 'any possible individual should admit of combination with itself' (Lewis 1986: 89) and argue that, given any actual iron sphere, there could have been two iron spheres having the same intrinsic pure properties of that actual iron sphere.

But, in fact, the actualist cannot appeal to the Recombination Principle. For the Recombination Principle is based on the 'Humean denial of necessary connections between *distinct* existences' (Lewis 1986: 87; my emphasis). That the principle is supposed to apply to *distinct* entities or existences is clear from Lewis' own gloss as the principle that 'anything can coexist with anything *else*' (Lewis 1986: 88; my emphasis). And, indeed, it is obvious that Lewis is right in restricting the principle to distinct existences, since the idea that anything can coexist with itself is a triviality that does not even deserve mention.

So, Lewis' assertion that any possible individual admits of combination with itself is misleading, since it suggests applying the Recombination Principle to a single individual. But what Lewis has in mind is the following: if there is a dragon in world $w1$, and there is a duplicate of that dragon in $w2$, then by duplicating each of those dragons the Recombination Principle guarantees that there is a world $w3$ containing two dragons—and since each dragon in $w3$ is a duplicate of the dragon in $w1$ (and of the dragon in $w2$), there is a sense in which every object can be combined with itself. That is, to obtain two duplicates of one and the same object in a single world by the Recombination Principle, there must already be some duplicate of the object.

In any case, the main point is that Lewis' Recombination Principle is based on Hume's dictum and this is meant to apply to *distinct* entities. But the actualist who is in the business of arguing against PIIb cannot assume that there are any distinct objects sharing all their intrinsic pure properties in the actual world. But then such an actualist cannot apply Lewis' Recombination Principle to 'generate' duplicates of any actual objects. Thus, the actualist cannot use Lewis' Recombination Principle to argue that, given any actual iron sphere, there could have been two iron spheres having the same intrinsic pure properties of that actual iron sphere.

4.4 Here is my argument against PIIb:

(1) It is metaphysically possible that at least one iron sphere has the following totality of intrinsic pure properties: *being made of iron, being spherical, having diameter D, having colour C, having temperature T, having mass M*, etc.

(2) If it is metaphysically possible that at least one iron sphere has the totality of intrinsic pure properties listed in premise (1), then it is metaphysically possible that two objects have those properties.
(3) Therefore, it is metaphysically possible that there are two iron spheres having the same totality of intrinsic pure properties.
(4) Therefore, PIIb is false.

The argument is valid. Are the premises of this argument true?

Premise (1) is true. For one could imaginatively conceive a situation in which there is at least one iron sphere of diameter D, colour C, temperature T, mass M, etc., and infer the possibility of such a situation from its imaginative conceivability. The content of such an act of imaginative conception is immune to Hacking-style objections: whether we are imagining one or two such spheres, we are in either case imagining at least one such sphere, and imagining that there is at least one sphere with such and such features is obviously very different from imagining that there is no such sphere.

One might think that there is another way of arguing for premise (1) that does not exploit any links between conceivability and possibility. This way consists in pointing out that there actually are iron spheres having a certain diameter, shape, colour, temperature, mass, etc., and then appealing to the axiom that whatever is actual is metaphysically possible. If so, one can just choose any actual iron sphere and let its diameter, shape, colour, temperature, mass, etc., be the diameter, shape, colour, temperature, mass, etc., specified in premise (1).

Unfortunately, this way of arguing for premise (1) might not give us what we need. This is because, for premise (1) not to beg the question against PIIb, the sphere in question must be such that none of its parts share all their intrinsic pure properties. There are many ways in which that might happen. For instance, the sphere could be such that no parts of it, however small, share exactly the same colour. Or it might be that no parts of it, however small, share exactly the same temperature. Or it might be divided in parts in such a way that no parts of it have exactly the same size and shape. And so on. Or it might be that although every intrinsic pure property had by one part of the sphere is had by another, there are no two parts of the sphere having all the same intrinsic pure properties. But I very much doubt that anyone has found an actual sphere satisfying such conditions. Nevertheless, it does not really matter, since we can imaginatively conceive such a sphere and infer its possibility from its imaginative conceivability. (This means, of course, that colour C or temperature T in premise (1) need not refer to a single homogeneous colour but perhaps to a distribution of colours or temperatures over the sphere. I shall ignore this complication in what follows).

Premise (2) is also true. First, note that one way in which premise (2) would be false would be if any or all of the properties listed in premise (1) were necessarily sufficient for being a certain object in particular. But the intrinsic pure properties of our iron sphere are not necessarily sufficient for being any object in particular. Indeed, some kinds are 'individuated', that is, defined, in terms of the intrinsic pure properties their members must have. And if a kind K is such that it can have many members, any possible member of K can have any set of intrinsic pure properties compatible with the intrinsic pure properties defining kind K (see Rodriguez-Pereyra 2017a: 3010 for a related principle). Such a kind is for instance the kind *iron*: it is individuated or defined by its atomic number, that is, the number of protons in the nucleus of every iron atom, namely 26. So any totality of intrinsic pure properties that a piece of iron can have is a totality of intrinsic pure properties that any piece of iron can have.[6] Thus, the intrinsic pure properties of our iron sphere are neither individually nor jointly necessarily sufficient for being any object in particular.

But that the intrinsic pure properties of our iron sphere are not necessarily sufficient for being any object in particular does not guarantee that it is metaphysically possible for two objects to share them. For it

---

[6] Does this mean that a frying pan made of iron could have been an iron sphere? To answer this we need to resolve the issue of whether the frying pan made of iron *is* a piece of iron. According to the monist view, it *is*, and so the frying pan could have been an iron sphere. This is, of course, consistent with the principle I am advocating above. Alternatively, one might adopt a pluralist view and maintain that the frying pan is *not* a piece of iron but is instead constituted by one. In that case, it is permissible to maintain that the frying pan could not have been an iron sphere. But this would not be a violation of my principle, since the frying pan, though made of iron, would not be a member of the kind *iron*, since more than being made of iron would enter into the definition of its kind. Thus, both monists and pluralists can accept my principle.

might be that, although more than one object could have those properties, no two objects could share them. Let us put this in the more graphic language of possible worlds: it might be that, although different objects have those properties in different possible worlds, there is no possible world where two objects have all those properties. And this might be because either those properties are jointly unshareable, or because at least one of them is individually unshareable. Is either of these possibilities the case?

The only properties that it is plausible to think are jointly unshareable are some sizes and shapes, for the size and shape of spacetime constrain the number, size, and shape of objects there could be (see Lewis 1986: 89). Given this, some sizes and shapes might be individually unshareable, since the size and shape of spacetime might be such that, necessarily, if an object has a certain size (shape) no other object has that size (shape)—in that case, even if having that size and shape is not necessarily sufficient for being any object in particular, no two objects could share such size or shape. But it might also be that although spacetime tolerates more than one object having a certain size, and it tolerates more than one object having a certain shape, it does not tolerate more than one object having both that size and shape. In this case, although that size and that shape are individually shareable, they are jointly unshareable. But, of course, there are sizes and shapes such that spacetime tolerates more than one object having *both* that size and shape, that is, there are sizes and shapes that are jointly shareable (and therefore they are individually shareable). And we may safely assume that the size and shape of spacetime tolerates two spherical objects of diameter D, the diameter mentioned in premise (1). For we can imaginatively conceive a sphere with a diameter of, say, one mile, since it is obviously possible that spacetime can accommodate two spheres of such a diameter, and let D be a diameter of one mile. Thus the shape and the size of the sphere mentioned in premise (1) can be safely assumed to be both jointly shareable.

Are any of the other intrinsic pure properties of our iron sphere individually unshareable? No. For intrinsic properties are such that having them is independent of how the rest of the world is. That means that, unless they are necessarily sufficient for being a certain object in particular, and unless the size and shape of spacetime prevents their shareability, any intrinsic pure property should be shareable. But we have seen that none of the intrinsic pure properties of our iron sphere are necessarily sufficient for being any object in particular, and only some sizes and shapes are individually or jointly unshareable due to the size and shape of spacetime. So, all the other intrinsic pure properties of our sphere are individually shareable. And we saw that we may safely assume that the size and shape of our sphere are jointly shareable, and therefore individually shareable. Thus, all the intrinsic pure properties of our iron sphere are individually shareable.

It should be clear that this is compatible with my conception of intrinsic pure properties (see Section 1.9), since intrinsic pure properties do not consist in being related, or failing to be related, to any external object at all. But other conceptions of intrinsic properties would agree with that result too. Just to give one example: Weatherson has as a necessary condition of intrinsic properties that they are independent not only of whether there are other objects, but independent also of what other types of objects there are (Weatherson 2001: 380; for Weatherson's restriction to pure properties, see Weatherson 2001: 367).

Thus, the intrinsic pure properties of our iron sphere are individually shareable. Could those properties be jointly unshareable? Sharing a group of properties is sharing each of them. So, if each property in that group is individually shareable, and the shape and size in that group are jointly shareable, all of them are jointly shareable, unless they are jointly necessarily sufficient for being a certain object in particular. But we have seen that it is safe to assume that the size and shape of our iron sphere are jointly shareable, and we have also seen that the intrinsic pure properties of our iron sphere are not jointly necessarily sufficient for being any object in particular. I conclude, therefore, that the intrinsic pure properties of our iron sphere are jointly shareable, and so it is metaphysically possible for two objects to share all of them. Therefore, premise (2) is true.

Thus, it follows from these true premises that it is metaphysically possible that there are two iron spheres having the same totality of intrinsic pure properties. This entails that PIIb is false, and therefore PIIb is false. Furthermore, Hacking's objection does not apply to the argument at any point.

4.5 The argument in the previous section establishes the metaphysical possibility of two iron spheres that are indiscernible with respect to their intrinsic pure properties. This shows that PIIb is false, but it does not show that PIIa is false, since it does not establish that the two spheres are indiscernible with respect to their extrinsic pure properties. For all the argument establishes, the spheres might be extrinsically different in virtue of how they are related to other objects, or even in virtue of how they are related to each other, if they happen to be related asymmetrically, say by one having come into existence before the other, or by one having impacted on the other and displaced it from its location, or by one having been a contributor to the production of the other (imagine one was made out of parts of the other), etc.

Now, clearly, it is metaphysically possible for two iron spheres to be only symmetrically related in so far as spatiotemporal and causal relations are concerned. For it is clearly possible for two iron spheres to be spatiotemporally symmetrically related and also causally symmetrically related (which would be the case, for instance, if they are never causally related at all). But since this possibility does not depend on how intrinsically similar the spheres are, if it is metaphysically possible that there are two iron spheres that share all their intrinsic pure properties, it is metaphysically possible that there are two iron spheres that share all their intrinsic pure properties and are spatiotemporally and causally symmetrically related with each other. I shall now give an argument that if it is metaphysically possible that there are two iron spheres that share all their intrinsic pure properties and are spatiotemporally and causally symmetrically related with each other, then it is metaphysically possible that there are two iron spheres that share all their pure properties, both intrinsic and extrinsic ones.

Before presenting the argument I need to introduce the notion of some objects being *independently* related by a relation R: some objects are independently related by a relation R if and only if their being R-related does not consist, even partly, in their being related in any way to any other concrete object.[7] Thus, when two objects are related by the

---

[7] As I said in Section 1.2 when I use 'object(s)' unqualifiedly, I mean concrete object(s). So the qualification here is not necessary. But it is anyway appropriate. This is because some might think that some relations between concrete objects consist in relations between those concrete objects and abstract objects. For instance, some might say that *being taller than* consists in two

relation of *being two miles away from*, they are independently related by that relation, but when they are related by the relation of *being equidistant from a planet*, they are not independently related by this other relation, since their being related by such a relation consists partly in their being at some distance from a planet.

The argument to come makes essential use of a version of a subtraction principle. I shall now state the strong version of the subtraction principle I have in mind, and then, when discussing one of the premises of the argument, I shall argue that only a weaker version is needed. The strong version says that if it is metaphysically possible that there are certain objects that have certain intrinsic pure properties and such objects are independently related to each other in certain ways, it is metaphysically possible for any of those objects and their parts to be the only objects that exist while having exactly the same intrinsic pure properties they had in the original situation and being independently related to each other in exactly the same way in which they were independently related in the original situation. This might be more clearly explained in terms of possible worlds: for every possible world w with more than one object, and for any objects xs having certain intrinsic pure properties in w and that are independently related to each other in some way in w, there is a possible world w^* whose only difference with w is that w^* contains just the objects xs and their parts, and where those objects have the same intrinsic pure properties they have in w and they are independently related to each other in the same way they are independently related to each other in w.

For ease of expression, I shall formulate the argument in terms of possible worlds, although this is not strictly necessary and it could be formulated using modal operators (indeed, as I have said several times already, I am not committed to *possibilism* here, so my appeal to possible worlds is *only* for ease of expression and should not be taken to commit me to *possibilia* of any kind). Here is the argument:

---

objects being related to a pair of heights such that the number corresponding to one of the heights is larger than the number corresponding to the other. But cases like these are not intended to be cases of concrete objects not being independently related. I am indebted to Erica Shumener for this point.

(1) There is a possible world *w* containing two iron spheres that share all their intrinsic pure properties and are spatiotemporally and causally symmetrically and independently related to each other.

(2) Either the two iron spheres are the only objects in *w* or there are more objects in *w*.

(3) If the two spheres are the only objects in *w*, then the spheres share all their pure properties, intrinsic and extrinsic.

(4) If there are more objects in *w*, then there is a possible world *w** containing only the two iron spheres and their parts, where these objects have the same intrinsic pure properties as they have in *w*, and where they are independently related to each other exactly as they are independently related in *w*—that is, in *w** the spheres, and their parts, share all their pure properties, intrinsic and extrinsic.

(5) Therefore, there is a possible world with two iron spheres sharing all their pure properties, intrinsic and extrinsic. That is, Black's world is possible.

(6) Therefore, PIIa is false.[8]

Since the argument is valid, the question is whether there is reason to believe in the premises. I have already argued, at the beginning of this section, that it is possible for two iron spheres sharing all their intrinsic pure properties to be spatiotemporally and causally symmetrically related. And, clearly, it is possible for two such iron spheres to be

---

[8] James van Cleve briefly sketched a similar argument, in the context of Modal Realism, against what he calls Weak PII, which roughly corresponds to my PIIa (he calls his operative principle 'Reverse Patchwork Principle' rather than 'Subtraction Principle': see van Cleve 2002: 390–1). But van Cleve's argument is conditional on the falsity of his Strong PII, which roughly corresponds to my PIIb, while I have argued against PIIb and so I can assume its falsity. Note also that the Subtraction Principle has been mainly used to argue for Metaphysical Nihilism, the view that it is possible that no concrete objects exist (there is a considerable literature on this, starting with Baldwin 1996 and Rodriguez-Pereyra 1997). The version of the Subtraction Principle stated above is more specific than the one used to prove Metaphysical Nihilism since to prove Metaphysical Nihilism one cannot restrict the principle to worlds containing more than one object and one does not need the clause that the objects in *w** have the same intrinsic pure properties they have in *w* and are independently related to each other in the same way they are independently related to each other in *w*.

spatiotemporally and causally independently related to each other. Thus, the first premise is true. And therefore the second premise is obviously true.

How about premise (3)? Since I am using 'object' as short for 'concrete object', that the spheres are the only objects in w is compatible with there being abstract objects in w. But, as I argued in Section 3.2, abstract objects do not make a difference to the pure properties of the spheres. Thus, premise (3) is true: a world in which the only (concrete) objects are the spheres is Black's world, a world in which the spheres share all their pure properties, intrinsic and extrinsic.

How about premise (4), the one where the Subtraction Principle is applied? Applying the Subtraction Principle to obtain world w^* presupposes that all other concrete objects in w are contingent entities such that (a) they are not necessitated by the spheres, (b) their non-existence does not necessitate the existence of anything else, and (c) their non-existence does not necessitate a difference in the intrinsic pure properties of the spheres or in the way the spheres are independently related to each other.

The only relatively controversial presupposition among these is that there are no necessary concrete objects in w. I am happy with that presupposition but, fortunately, I do not need to assume it. For, as I said in Section 3.3, the only allegedly necessary object that is even remotely plausible to believe in is God. And I argued in that section that it is not the case that God must be differently related to the spheres in every world in which the only other concrete objects are two spheres.

Thus, the mere supposition of a necessary concrete object does not entail the impossibility of the spheres sharing all their pure properties. To make it impossible that the spheres share all their pure properties because of necessary concrete objects one needs to suppose that there are necessary concrete objects that must be differently related to the spheres. But there is no reason at all to believe in such necessary concrete objects.

In any case, if there are any necessary objects, whether abstract or concrete, the Subtraction Principle cannot 'generate' a world in which there are only two iron spheres, since in that case there is no such world. But then one only needs to restrict the Subtraction Principle to contingent concrete objects, so that it requires that it is metaphysically possible for any *contingent* objects to be the only *contingent* objects that exist while having exactly the same intrinsic pure properties they had in the original world and being independently related exactly in the same way in which they were independently related in the original world.

But some will object even to this weakened version of the principle, since they believe that origins are metaphysically necessary (see Kripke 1981: 112–13), and therefore an origin cannot be subtracted from a world without subtracting what it originates. I do not believe in the metaphysical necessity of origin, but I can accommodate those who believe in it by weakening the Subtraction Principle even further. Indeed, the argument will go through with a version of the principle that only requires that it is metaphysically possible for *some* contingent objects to be the only contingent objects that exist while having exactly the same intrinsic pure properties they had in the original world and being independently related exactly in the same way in which they were independently related in the original world. This allows that 'subtracting' some objects—say, origins—from a world entails 'subtracting' other objects—say, what those origins originate—from it.

But is it plausible that iron spheres satisfy this latter version of the Subtraction Principle? I think it is, even under the assumption that origins are metaphysically necessary, for it is metaphysically possible for iron spheres not to have any causal origin at all.

But one might even agree that the spheres in question have a causal origin. For if it is possible for two iron spheres to have the same causal origin and be symmetrically related to their causal origin, and if it is possible for the causal origin not to originate anything other than the spheres, the version of the Subtraction Principle mentioned in the previous paragraph would give us a world where only the spheres and their causal origin exist, and where the spheres have exactly the same intrinsic pure properties they had in the original world, they are independently related to each other in the same way in which they were independently related in the original world, and they and their causal origin are independently related to each other in the same way in which they were independently related in the original world.[9] Of course, this

---

[9] Note that nothing here entails that in the resulting world the spheres must be only independently related to each other—indeed, they will not be, since they will bear to each other the

world is not Black's world, since it contains an object other than the two spheres and their parts. But this is irrelevant: such a world falsifies PIIa no less than Black's world does.

It is important to note that the parts of the spheres do not present a threat to the argument. For given that the spheres share their colour, shape, temperature, mass, diameter, etc., to any given part x of one of the spheres there must correspond a part y of the other sphere that has exactly the same pure properties, both intrinsic and extrinsic, as x has. Indeed, if this were not the case, the spheres would not have exactly the same pure properties: for example, one would have a pure property like *having a part having pure properties F, G, and H*, while the other sphere would lack this property.

But don't the spheres compose another concrete object of which they are parts? And don't the parts of one of the spheres compose objects with the parts of the other sphere? This would be the case, for instance, if Universal Composition, the principle that any non-overlapping objects compose another one, were true. In that case, of course, Black's world would not be possible: the spheres and their parts could not exist alone—indeed, they could not even be the only *contingent* existing objects. One option would be to reject Universal Composition. But that, by itself, does not guarantee that it is possible for the spheres and their parts to exist alone, since that leaves open the possibility that *some* of those objects must compose something else. One way to guarantee that Black's world is possible would be to postulate that spatiotemporally separated objects do not compose any objects. But that is not necessary. All one needs to assume is that iron spheres like those in Black's world do not compose any object (and that parts belonging to different spatiotemporally separated iron spheres like those in Black's world do not compose any objects). And that is a plausible assumption.

Now, some people believe in Universal Composition, and so they cannot accept the possibility of Black's world. But it doesn't really matter.

---

relation of *originating in the same object*, which partly consists in their being related in a certain way to a third object. But this does not matter. What matters is that in the resulting world the way in which the spheres are independently related to each other is the same way in which they were independently related to each other in the original world, and the way in which they and their causal origin are independently related to each other is the same way in which they were independently related to each other in the original world.

For those extra objects they believe in will not make a difference to the pure properties of the spheres. Indeed, the spheres are symmetrically related to some of those objects and, for every object such that only one of the spheres is related to by a relation R, there is another object with the same pure properties to which the other sphere is related to by R (if that were not the case, there would be a part x of one of the spheres such that no part of the other sphere has the same pure properties as x, which would mean the spheres do not have the same pure properties).

But what if some parts of sphere a compose something with some parts of sphere b (or with sphere b itself), but the corresponding parts of sphere b do not compose anything with the corresponding parts of sphere a (or with sphere a itself)? If that were the case, the spheres would not be indiscernible with respect to their pure properties, since one sphere would have a certain pure property (for instance, sphere a would have the property of *composing an object with an atom of another sphere*), while the other sphere would lack such a property. But the view that some parts of sphere a compose something with some parts of sphere b (or with sphere b itself), but the corresponding parts of sphere b do not compose anything with the corresponding parts of sphere a (or with sphere a itself) is absurd. For it offends against the extremely plausible principle that composition supervenes upon the pure properties and relations of the component objects—that is, the principle that, necessarily, no objects xs compose another one unless any other objects ys sharing their pure properties and relations with the xs compose something too. This is, indeed, extremely plausible, and it should be noted that this principle does not conflict with the view that composition is brutal—that is, the view that there is no true, non-trivial, and finite answer to the Special Composition Question, the question of the necessary and sufficient conditions for composition (Markosian 1998: 214).[10] Thus, nothing here commits me to the rejection of the view that compositional facts are brute.

---

[10] Ned Markosian, the main defender of the brutality of composition and compositional facts, would surely agree with this. For he thinks that the brutality of composition and compositional facts is consistent with the *global* supervenience of composition upon 'non-mereological universals' (Markosian 1998: 216). And although his global supervenience thesis does not entail my supervenience thesis, which is formulated in terms of a plural weak notion of supervenience, they are both in the same spirit.

Now, although we can deny that the spheres compose another object and that the parts of different iron spheres compose other objects, we cannot deny that each sphere belongs to certain sets the other sphere does not belong to. And perhaps we should accept that each sphere is a constituent of certain states of affairs that the other one is not a constituent of. But it should be clear now that such differences will not make a difference to the pure properties of the spheres. Both spheres will share properties like *being a member of a singleton set whose member is an iron sphere*, *being a constituent of a state of affairs involving an iron sphere*, and so on, and the properties they will not share will be impure properties like *being a member of {a}*, *being a constituent of a state of affairs involving iron sphere a*, and so on.

There is another way in which premise (4) might be resisted. For it might be argued—better: one might *suppose* that it might be argued—that it is metaphysically necessary that the number of contingent spatiotemporally separated objects is higher than 2. For instance, it might be argued that there must be at least three, or four, or fifty, or infinitely many contingent spatiotemporally separated objects. I believe any of these theses to be false, since I believe that there could have been no concrete objects (Rodriguez-Pereyra 1997). Nevertheless, even if any of these theses are true, it is clear that it is metaphysically possible that the only contingent spatiotemporally separated objects are, for instance, iron spheres. Now, if n (where n is any positive integer higher than 2) is the minimum number of contingent spatiotemporally separated objects that can exist, it is also metaphysically possible that there are n iron spheres arranged in such a way that they share all their pure properties, intrinsic and extrinsic, if they are the only objects in the world: for this condition to obtain all that has to be the case is that the spheres share all their intrinsic pure properties and are arranged in the shape of an n-sided regular polygon. And if there must exist infinitely many contingent spatiotemporally separated objects, let an infinity of iron spheres with the same intrinsic pure properties be arranged in a straight line with no first or last sphere: in that case they will also share their extrinsic pure properties. So, those who think that they have an argument that the minimum number of contingent spatiotemporally separated objects that can exist is higher than 2, can use the argument above except that, instead of making reference to two intrinsically indiscernible spheres, they will have to refer to n iron spheres arranged in the shape of an n-sided regular polygon or to an infinity of spheres arranged in a straight line with no first or last sphere—the argument will then deliver the same conclusion as before: PIIa is false. (Note that nothing requires that the objects in question be all iron spheres: if, for instance, $n = 3$, a world with only two iron spheres and a perfectly symmetrical banana symmetrically related to the spheres would also falsify PIIa).

Finally, I would like to consider a version of Hacking's objection to my argument. The idea behind premise (4) is simply that if the two spheres could have had the same intrinsic pure properties while existing alongside other objects, they could have existed with those same intrinsic pure properties while nothing else exists and they are only symmetrically related. But someone might adopt a counterpart theoretic semantics and maintain that to say that the two spheres could have been the only existent objects is to say that there is a possible world w^* in which each sphere has a counterpart and nothing else has a counterpart. But, on Counterpart Theory, it does not follow from this that each sphere has a different counterpart in w^*, for on Counterpart Theory different objects can have the same counterpart at one and the same possible world. Therefore, simply pointing out that the two spheres could have been the only existent objects does not establish that there is a possible world where only two spheres exist. That is, the objection rejects the assumption that subtracting every other object from a world in which the spheres share all their intrinsic pure properties and are only symmetrically related to each other delivers a world with only two spheres in it. But without such a world one has not established line (5) of the argument.

Now, as I have said, my reference to possible worlds was only for ease of expression, since I believe in Actualism. Nevertheless, although Counterpart Theory was originally formulated to account for *de re* modality in the context of Lewisian Modal Realism, Counterpart Theory can be adapted to an actualist setting. For the main idea of Counterpart Theory is that the truth conditions of *de re* modal statements are given in terms of the counterparts of the objects the statements are about, and counterparts need not be interpreted as merely possible, non-actual objects (see, among others, McMichael 1983, Lewis 1986: 237–9, and Wang 2015). So, I cannot meet the objection simply by switching to an actualist setting.

But the objection can be met. For Counterpart Theory can and should be made to assign counterparts not only to individual objects but also to tuples of objects (as noted in Hazen 1979: 334. See also Lewis 1983c: 44–5 and Lewis 1986: 232–4). And, clearly, an individual sphere, although it may be a counterpart of both spheres *a* and *b*, is not, on any reasonable counterpart relation, a counterpart of the *pair* of spheres *a* and *b*. Indeed, on any reasonable counterpart relation, only a pair of spheres is a counterpart of the pair of spheres *a* and *b*. Thus, subtracting every other object from a world containing two spheres delivers a world with only two spheres in it, even on Counterpart Theory. Thus even those with leanings towards Counterpart Theory should accept my argument.

Thus, premise (4) is true—or, if it is not true as stated, there are easy ways of rewriting it so that it is true and it ensures the conclusion of the argument. Since the other premises are true, and the argument is valid, the conclusion is established: PIIa is false. As I have said at various points, I am happy with the relatively controversial assumptions of the argument and therefore I take the argument to establish not only the falsity of PIIa but also the possibility of Black's world. But I have also shown that those who reject those presuppositions have ways of reformulating premise (4) in such a way that, although Black's world may not in that case be possible, the argument still leads to the conclusion that PIIa is false. And that is the really important conclusion.

There are two things I need to note about my argument against PIIa. The first is that it presupposes the falsity of PIIb, and that is why I first argued for the falsity of PIIb. The second is that it is not vulnerable to Hacking's objection. As we just saw, premise (4) is not vulnerable to Hacking's objection. The other premise that could be thought to be subject to Hacking's objection would be premise (1). But that premise is secured by the argument from the previous section, which establishes that it is metaphysically possible that there are two iron spheres sharing all their intrinsic pure properties, plus the consideration that if it is metaphysically possible that there are two iron spheres sharing all their intrinsic pure properties, it is metaphysically possible that there are two iron spheres sharing all their intrinsic pure properties that are spatiotemporally and causally symmetrically and independently related to each other. But neither the argument from the previous section, nor this other consideration, is vulnerable to Hacking's objection. Therefore, the argument that PIIa is false is immune to Hacking's objection.

4.6 Some have argued that one way of defending PIIa in the face of Black's world is to adopt the 'summing defence', which consists in saying that Black's world contains, not two spheres, but one simple and scattered object occupying two spherical regions (Hawley 2009: 106). This object is not a sphere nor does it have any spheres as parts. If that is the case, Black's world does not violate PIIa.

The summing defence might have some bite if presented as an objection to an argument for Black's world based on its imaginative conceivability. In that case, the objector could say that there is no guarantee that what we are conceiving when we imaginatively conceive Black's world is a world containing two spheres as opposed to one simple but scattered object occupying two spherical regions. But what my argument establishes is precisely the possibility of a world containing two spheres that share all their pure properties.

Indeed, the summing defence requires rejecting the Subtraction Principle. For according to the Subtraction Principle, if there are any (contingent) objects, any subgroup of them can exist without the others. But the summing defence requires that if there are some objects that include two iron spheres that share all their intrinsic pure properties and are spatiotemporally and causally symmetrically and independently related to each other, these spheres cannot exist on their own while keeping the same properties and relations to each other—that is, the other objects cannot be 'subtracted' without either eliminating or altering the spheres. And this is absurd, indeed, incredible.

Hawley distinguishes two other ways of defending PIIa from Black's world (Hawley 2009: 106). One is to argue that there are pure properties with respect to which the two spheres must differ. But I have already argued against such a possibility in Chapter 3. The other defence is to argue that what Black's world contains is a single, bi-located sphere. This defence has no chance given my argument for the possibility of Black's world, which establishes that it contains *two* spheres—and, as we saw, the argument is not vulnerable to Hacking's objection, which is how this defence gets articulated as an objection to an argument for Black's world based on its imaginative conceivability.

But there is yet another defence of PIIa from Black's world. This, the so-called *overlap defence*, consists in arguing that the spheres (and objects in general) are fusions of pure properties. In that case, the spheres share among their 'qualitative parts' an unlocated sphere that consists of all their pure properties except their location properties. So, according to this line of thought, in Black's world there is one unlocated sphere but two located ones, and the two located ones differ with respect to their location properties (Shiver 2014: 909–10). But there are two problems with this defence. The first is that it presupposes that properties are concrete objects—or else that spheres are abstract objects! For no fusion of abstract objects can be a concrete object, if one understands concrete objects as those that exist in space, time, or spacetime and abstract ones as those that do not; and this is how I am understanding these terms (see Section 1.2). But properties, as I argued in Section 1.4, are abstract objects, and iron spheres are concrete ones. The second problem is that the location properties that enter the fusions that make up the located spheres must be pure properties, but as we saw in Section 3.3, the spheres share all their pure location properties, and therefore they do not differ with respect to them.

Anthony Shiver, the proponent of this defence, is aware of this problem and his answer is to treat location properties as brutally distinguished individuals, and treat occupation as the fusion of location properties with other properties (Shiver 2014: 912). But this I find unintelligible. Location properties are either relational properties such that having them consists in bearing the occupying relation to a region or regions, or else they are relational properties such that having them consists in bearing distance relations to other objects. And, again, understood in either of these ways, the spheres in Black's world only differ with respect to their *impure* location properties.

4.7 The argument presented against PIIa in Section 4.5 is a subtraction argument, and it presupposes the falsity of PIIb. The next question, then, is whether some other version of the subtraction argument can be used to argue against PIIb. I can think of one subtraction argument against PIIb and, in my view, it fails.

The subtraction argument I have in mind starts with a possible world in which, say, two iron spheres are intrinsically purely qualitatively indiscernible except for the fact that one of them has an extra part, say an extra part of matter in its surface. Subtracting that extra part of matter—which may be as large as you please (see Rodriguez-Pereyra 2017a: 3016)—takes us into a possible world in which the two spheres share all their intrinsic pure properties. The problem with this argument is that it presupposes the falsity of PIIb. Indeed, in the initial world there are two spheres, *a* and *b*, and one of them, let it be *a*, has an extra part of matter, call it *c*; and so in that world there are two objects that share all their intrinsic pure properties: namely *b* and a part of *a*, *a minus c*. Therefore, this subtraction argument cannot establish the falsity of PIIb, because it presupposes its falsity.

One might think that there is another version of the subtraction argument against PIIb in which one subtracts properties from objects. On this version we start with a possible world in which one object, say an iron sphere, has one or more intrinsic pure properties than the other one. Subtracting the additional intrinsic pure properties from that sphere is supposed to take us into a possible world in which the two spheres share all their intrinsic pure properties.

But there is no such subtraction of properties. For properties are predicable conditions, and such conditions exist whether or not anything satisfies them. Indeed, such conditions are necessarily existent abstract objects. Therefore, it is impossible to subtract any properties.

One might think that this is a misinterpretation of the argument: what the argument requires is that we subtract a property *from a sphere*, not that we subtract the property in question *from the world*. According to this, what gets subtracted from the world is not really a property, but a fact: the fact that a certain sphere has a certain property.

But there is no such subtraction of facts. Indeed, it is essential to the subtraction argument that the objects subtracted from a world are not replaced by any others in the resulting world. But on my conception of properties, there are negative properties, and if any object lacks a certain property, it has the corresponding negative property. Thus, if the fact that a certain sphere has a certain property fails to obtain, the fact that the sphere in question has the corresponding negative property must obtain. Therefore, going from a world in which one sphere has a certain property to a world in which it lacks it, does not involve merely subtracting a fact from the first world: it involves replacing it by another one in the resulting world. Thus the argument in question is not a version of the subtraction argument. Indeed, one condition of possibility of *object* subtraction arguments is precisely that the non-existence of the subtracted objects does not necessitate the existence of any other object. But the analogous condition for facts does not hold: the non-obtaining of any fact necessitates the obtaining of another one.

It might be said that it does not really matter whether the argument is a subtraction argument or not, what matters is that it establishes the falsity of PIIb. But, contrary to this, what is relevant here is whether the argument is a version of the subtraction argument. For I have already given an argument that refutes PIIb, and since the fact that there is such an argument obviously does not exclude the possibility of there being other viable arguments that refute PIIb, the existence of alternative arguments that refute PIIb is not relevant for me. But since I have offered a subtraction argument against PIIa, the relevant question is the one I set up to investigate, namely whether there is a subtraction argument that refutes PIIb. For, given that my argument against PIIa presupposes the falsity of PIIb, if there is a subtraction argument against PIIb, using both subtraction arguments, one against PIIb and the other against PIIa, would contribute to the argumentative unity of my case against PIIa. But I have not found any successful subtraction argument against PIIb.

5
Two Arguments for PII

5.1 There are two ways or methods of proving or establishing PII or any of its versions. One way is to find a relevant kind of property such that (a) every concrete object must have one property of that kind and (b) no two concrete objects can share properties of that kind. This method establishes PII by showing that every concrete object must have an unshareable property of the relevant kind. If one successfully established, for instance, that concrete objects must have pure individual essences one would have established PIIa, and therefore PII too, by this method.

The other method does not appeal to any unshareable properties. Instead, if successful, it shows that concrete objects cannot share all their properties of the relevant kind, but without appealing to any unshareable properties. Leibniz's argument for PII in his Correspondence with Clarke is an instance of this method of argument. In that argument Leibniz does not appeal to any unshareable properties, but he argues that the possibility of two objects' sharing all their properties violates the Principle of Sufficient Reason, which he takes to be true (Leibniz 1956: 61).

I shall now present two arguments for PII, the first one will be of the second type, and the second one of the first type.

5.2 Before presenting the first argument, and by way of preparation for it, let us note that NT-indiscernible objects are incompatible with Hume's dictum, according to which there are no necessary connections between numerically distinct concrete objects. Understanding necessary connections in a minimal way, we can formulate Hume's dictum as follows:

Hume's dictum: Necessarily, no object necessitates another.

Now, if two objects share all their NT-properties, they necessarily coexist. Indeed, any object necessarily coexists with itself. Therefore, object *a* has the property of *necessarily coexisting with a*. The property of *necessarily coexisting with a* is an NT-property (if two objects differ with respect to it, they must differ not only numerically but also in how they coexist with *a*). So, anything sharing all its NT-properties with *a* also has the property of *necessarily coexisting with a*, and therefore it necessarily coexists with *a*. But any object that necessarily coexists with *a* necessitates *a*'s existence, and *a* necessitates the existence of anything that necessarily coexists with it.

Thus, one could argue for PII from Hume's dictum. But such an argument would have severe limitations, since Hume's dictum, as I have stated it, is either implausibly strong, or too weak to play the role it needs to play in the argument. If by distinct entities it is meant numerically distinct entities, then Hume's dictum is implausibly strong since, as it has been pointed out by Jessica Wilson, there are counterexamples to it—for example, the existence of a states of affairs necessitates that of its constituent parts (Wilson 2010: 601).

One possible response to this limitation would be to weaken Hume's dictum and make it apply only to objects that are mereologically distinct. But there are two problems with this. One is that there are good reasons to think that Hume's dictum should apply to all numerically distinct objects, not just to mereologically distinct objects (see Bohn 2014: 345–50). The other problem, specific to our discussion, is that there is no restriction in PII to mereologically distinct objects. Thus such a weakened version of Hume's dictum would not be strong enough to deliver PII.

A better response to the limitation arises from noting that, since objects sharing all their NT-properties would necessarily *coexist*, they would *mutually* necessitate each other. For however plausible is the idea that there are numerically distinct objects that necessitate others, it is less plausible to suppose that there are numerically distinct *concrete* objects that *mutually* necessitate each other. Thus, concrete objects sharing all their NT-properties would offend not only against Hume's dictum, but also against a weaker, and therefore more plausible principle, to the effect that no two concrete objects can be *mutually* existentially dependent. Let us formulate this version of Hume's dictum, ruling out mutual necessitation between concrete objects, as follows:

HDM: Necessarily, no two objects mutually necessitate each other.

This principle is rather plausible. One indication of its plausibility is the traditional conception of substance (substances being paradigmatic concrete objects) as an independent entity. This tradition, going back to Aristotle and passing through such eminences as Descartes, does not form a uniform, monolithic body of thought on substance, but is coloured by a variety of different interpretations of the notion and modal strength of independence. Nevertheless, despite these variations, such a tradition will, almost if not completely unanimously, agree that necessarily coexistent objects are not substances. And, in fact, Bernard Katz argued for a version of PII restricted to substances from the idea that substances cannot necessarily coexist (Katz 1983: 42–4).

True, the independence conception of substance is not the only conception of substance, and it has its detractors. But the idea that there cannot be mutual necessary existential connections between any two concrete objects is independent of the independence conception of substance. Bob Hale, for instance, adheres to it when he claims that, if *a* and *b* are concrete, it is possible that one of them should exist without the other (Hale 2015: 187).[1] Furthermore, typical counterexamples to the independence conception of substance are not counterexamples to the idea that there cannot be objects that share all their NT-properties. For such counterexamples typically involve entities (allegedly substances) in one-way relations of necessitation to other entities, which means that those entities do not share all their NT-properties. And even when they involve entities in mutual relations of necessitation, such entities fail to share all their NT-properties—for instance, some might allege that

---

[1] Hale endorses a principle trivially logically equivalent to the claim that if, necessarily, two things share all their properties, they are numerically identical (Hale 2015: 188). But this is not my PII, nor does it entail my PII, since he is restricting his quantifier to purely general properties (Hale 2015: 166), which are my pure properties. Thus, what he is committed to is similar to PIIa, but it does not entail it either, since in PIIa the necessity operator modifies the conditional, while in Hale's principle it modifies the antecedent of the conditional.

forms necessitate hylomorphic compounds as much as these compounds necessitate forms (see Koslicki 2018: 189 for more on this case), but forms and hylomorphic compounds do not share all their NT-properties: the former have the property of *being a form* and the latter do not. Or consider Kit Fine's necessarily coincident letters (Fine 2000: 359–60). These letters mutually necessitate each other, but they do not share all their NT-properties: one is written in Prittle, the other one in Prattle, one is addressed to a person, the other one is addressed to another person, and so on.

Another indication of the plausibility of HDM is that plausible examples of mutually necessitating entities typically involve abstract objects: sets, numbers (assuming, for the sake of increasing our list of examples, that they are not sets), properties and propositions (on conceptions of properties and propositions according to which they are abstract objects), points and regions of space, times and periods of time, abstract geometrical figures, and so on. Note that I am not saying that abstract objects can only mutually necessitate other abstract objects: Socrates and {Socrates} necessitate each other, for instance. But, typically, in plausible cases of mutual necessitation at least one of the objects involved is abstract. We are inclined to allow mutual necessary connections in these cases because the necessary coexistence of the objects in question seems explainable and therefore understandable. In at least some cases, the explanation is simply that abstract objects are necessary objects, that is, that they exist necessarily. This is the reason why number 8 and The Triangle necessarily coexist, since no two necessary objects can fail to necessarily coexist (I am, of course, assuming that numbers and geometrical figures are necessary objects, an assumption which, although plausible, can be rejected: see Cowling 2017: 201–6 for discussion).

Shouldn't then HDM be restricted to *contingent* concrete objects? For two necessary concrete objects—if a multiplicity of necessary concrete objects were possible—would have to necessarily coexist, since their necessary coexistence is simply a consequence of the fact that each one of them is a necessary object. But restricting HDM to contingent concrete objects would rob the principle from its role in establishing PII, since there is no restriction to contingent objects in PII: if there were necessary concrete objects that shared all their NT-properties, they would falsify PII.

Now, as I indicated in Section 3.3, I do not believe in any necessary concrete objects. Since the idea of a necessary concrete object is controversial, and much more so is the idea of a multiplicity of necessary concrete objects, why not restrict PII to contingent objects? There are two reasons not to do so. One is that although controversial, the idea of a multiplicity of necessary concrete objects does not seem to be incoherent, and so the restriction would detract from the interest of PII by decreasing its generality.

The other reason is that, however plausible HDM seems, there are some versions of plenitude that entail that even contingent concrete objects can mutually necessitate each other, as Alex Roberts has pointed out to me. Take a version of the plenitude principle according to which every region of spacetime that contains an object contains objects instantiating every possible modal profile (cf. Bennett 2004: 354). Take any two contingently coexisting objects a and b. By the version of the principle of plenitude just mentioned there will be, exactly where a is, an extremely fragile object otherwise exactly like a except that it could only exist in the precise conditions obtaining in the actual world—had the world been very slightly different in any way, that object would not have existed. Call that object a^*. Now, by the same version of the principle of plenitude there will be, exactly where b is, an extremely fragile object otherwise exactly like b except that it could only exist in the precise conditions obtaining in the actual world—had the world been very slightly different in any way, that object would not have existed. Let us call that object b^*. a^* and b^* are contingent concrete objects, and yet they necessarily coexist, and therefore they mutually necessitate each other. Now, I want to remain neutral about plenitude, so I am not presenting this as a refutation of HDM; my point is simply that there is a coherent way to reject HDM, a way that may have a few supporters who will not therefore be willing to accept PII on the basis of HDM.

But it is not necessary to appeal to HDM in order to support PII, since there is a more plausible principle which rules out both necessary and contingent concrete objects that share all their NT-properties. The general idea behind this other principle is that no two concrete objects can be mutually *descriptively* dependent.

What is it to be mutually descriptively dependent? Let me introduce some terminology and say that two objects x and y necessarily co-vary with respect to a property F if and only if, necessarily, x has F if and only if y has F. Then we can formulate the principle of no mutual descriptive dependence for concrete objects, NMDD, as follows:

NMDD: Necessarily, no two objects necessarily co-vary with respect to every NT-property.

In terms of possible worlds, what NMDD says is that in every possible world every two concrete objects are such that, in some possible world, they differ with respect to some NT-property.

Since necessarily coexistent objects necessarily co-vary with respect to some NT-properties, for instance the property of *existing*, HDM entails NMDD. But NMDD does not entail HDM. Indeed, unlike HDM, NMDD is consistent with a multiplicity of necessarily existent objects, for NMDD rules out only those necessary objects that necessarily co-vary with respect to all NT-properties. Thus, NMDD is more plausible than HDM.

Indeed, it seems to me that NMDD is plainly true. For is it not patently false that two concrete objects a and b could be such that, necessarily, a is red if and only if b is red, a is yellow if and only if b is yellow, a is square if and only if b is square, a is taller than the tallest man if and only if b is taller than the tallest man, a is next to the Statue of Liberty if and only if b is next to the Statue of Liberty, a caused c if and only if b caused c, a is human if and only if b is human, a is exactly in place p at time t if and only if b is exactly in place p at time t, and so on? Indeed, it is patently false that concrete objects can have such a degree of mutual dependence—any objects exhibiting such a degree of mutual dependence would have very little degree of ontological descriptive autonomy, less ontological descriptive autonomy than is required for anything to be a concrete object—they would be ontological shadows, indeed, mutual shadows: shadows of a shadow.[2]

---

[2] Yablo advances a superficially similar but actually very different and stronger thesis than NMDD. His thesis is that no distinct objects can have the same *categorical* properties in the

Not only can one patently see that the supposition of two concrete objects exhibiting such a degree of mutual descriptive dependence is false, one can also understand why it is false. If we restrict our consideration to properties that are contingent to objects, the properties of a concrete object depend on those of another when those objects are connected either causally or constitutively.

Take the latter case first. Objects are connected constitutively when one is a part of another on a broad understanding of the notion of part. Even if the part–whole relation produces dependence between the properties of the part and the whole, no proper part and its whole will necessarily co-vary with respect to all the NT-properties, since a property like *being a proper part of b* is an NT-property, and it is not the case that, necessarily, a is a proper part of b if and only if b is a proper part of b. There is another reason why the part–whole relation does not induce the degree of descriptive dependence required by necessary co-variation with respect to all NT-properties. For even if the part–whole relation induces any necessary connections at all, it is plausible that such connections are not mutual. Indeed, if parts are necessary for their wholes to exist, it is plausibly not the case that wholes are necessary for their parts to exist, and if wholes are necessary for their parts to exist, it is plausibly not the case that parts are necessary for their wholes to exist. But if this is the case, then the properties of parts and wholes do not necessarily co-vary, since it would be possible for a part or a whole x to

---

same possible worlds (Yablo 1987: 311), where categorical properties are those that are grounded in how an object happens to be rather than in how the object could be or would be were the circumstances different (Yablo 1987: 305). In my terminology, this means that no numerically different objects can necessarily co-vary with respect to all their *categorical* properties—but not all NT-properties are categorical, which is why Yablo's thesis is stronger than NMDD. Furthermore, he does not support his thesis by arguing that such necessary co-variation would destroy the descriptive autonomy required of concrete objects. Instead, he supports his thesis by rejecting the possibility that two objects having the same categorical properties, say two spheres, could be located in exactly the same place. Indeed, he says that such a hypothesis is not only beyond our powers of belief but also beyond our powers of stipulation (Yablo 1987: 312), which implies that we cannot even make sense of two spheres being located in the same place. But the hypotheses that there might be two spheres in the same place *makes perfect sense*. Furthermore, not everything need be in space, so it is not clear how Yablo would rule out the possibility of two unlocated Cartesian minds necessarily co-varying with respect to their categorical properties—true, Yablo seems to have found no reason to believe in Cartesian minds (Yablo 1990: 200), but even if Cartesian minds do not actually exist, what is relevant to PII is their possibility.

have the property of *existing* without the corresponding whole or part y existing, in which case x and y would not necessarily co-vary with respect to existence.

How about causal connections? Can they generate the degree of descriptive dependence required by necessary co-variation with respect to all NT-properties? Presumably, if two objects necessarily co-vary with respect to all NT-properties due to a causal connection this is because their having the NT-properties they have is due to a common cause. Thus, for every NT-property F had by a and b, there is an object c that causes both a and b to have F. But the two occurrences of the necessity operator in NMDD express metaphysical necessity, and thus even if for every NT-property F had by a and b there is an object c that causes both a and b to have F, this would not violate NMDD, provided the sphere of what is metaphysically possible exceeds that of what is causally possible, a supposition I cannot defend here, but which I think is extremely plausible.

The idea that mutual causation could account for necessary co-variation with respect to all NT-properties does not fare well either. For if two objects a and b are mutual causes, and such mutual causation accounts for their necessary co-variation with respect to all NT-properties, then for every property F a has, it has the property of *having F as an effect of b's causal action*, and similarly, for every property F b has, it has the property of *having F as an effect of a's causal action*. But since these are NT-properties, and they are supposed to share all their NT-properties, it follows that for every property F a has, it has the property of *having F as an effect of a's causal action*, and for every property F b has, it has the property of *having F as an effect of b's causal action*. But this kind of self-causation is absurd.

Thus, I take NMDD to be true. But if NMDD is true, there cannot be objects that share all their NT-properties—that is, if NMDD is true, PII is true. The argument that establishes this is simple:

(1) Necessarily, if two objects share all their NT-properties, they necessarily co-vary with respect to every NT-property.
(2) Necessarily, no two objects necessarily co-vary with respect to every NT-property (NMDD).
(3) Therefore, necessarily, no two objects share all their NT-properties (PII).

The argument is valid and simple. And I have already argued for the second premise, which is NMDD. But it is important to point out that NMDD is distinct, and logically weaker, than PII. This may be more clearly seen if the theses are put in terms of possible worlds. NMDD says that in every possible world every two concrete objects are such that, in some possible world, they differ with respect to some NT-property. PII says that in every possible world every two concrete objects differ with respect to some NT-property. Thus, while PII entails NMDD, NMDD does not entail PII.

How about premise (1)? It is necessary that every object *x* is such that, for every property F, necessarily, *x* has F if and only if *x* has F. So, it is necessary that every object *x* is such that, for every property F, it has the property of *necessarily, having F if and only if x has F*. Therefore, this is true of every NT-property. Thus, take any object *a*: it is necessary that *a* has, for every NT-property F, the property of *necessarily, having F if and only if a has F*.

Now, if F is an NT-property, *necessarily, having F if and only if a has F* is also an NT-property. For an NT-property is one such that either it is not possible that two objects differ with respect to it, or differing with respect to it requires differing extra-numerically. Suppose first that it is possible that objects differ with respect to F. Then it is possible that objects differ with respect to the property of *necessarily, having F if and only if a has F*. Indeed, any objects differing from *a* with respect to F will differ from *a* with respect to the property of *necessarily, having F if and only if a has F*. Suppose now that it is impossible that objects differ with respect to F. Even so, any object that does not necessarily coexist with *a* will differ from *a*, or from any object that necessarily coexists with *a*, with respect to the property of *necessarily, having F if and only if a has F*. And it is a very plausible assumption that some objects fail to necessarily coexist with each other (which entails, of course, that theses like that all actual objects exist necessarily are very plausibly false). Thus, whatever property F is, the property of *necessarily, having F if and only if a has F* is such that it is possible for two objects to differ with respect to it.

Thus, if F is an NT-property, for the property of *necessarily, having F if and only if a has F* to be an NT-property it must be the case that differing with respect to it requires differing extra-numerically. But differing with respect to the property of *necessarily, having F if and only if a has F* requires differing extra-numerically, since it requires differing with respect to whether one's having F is necessarily connected to *a*'s having F.

Therefore, if F is an NT-property, so is the property of *necessarily, having F if and only if a has F*. But then, if *a* and *b* share all their NT-properties, for every NT-property F, *b* has the property of *necessarily, having F if and only if a has F*. Thus, if *a* and *b* share all their NT-properties, they necessarily co-vary with respect to all the NT-properties, which is what premise (1) asserts. Thus, the argument is valid and its premises are true. It therefore establishes that there cannot be objects that share all their NT-properties, that is, it establishes PII. In the next section I shall discuss some aspects of the argument.

5.3 First, it is important to note that necessarily co-varying with respect to a property F does not entail possibly having F. Premise (1) of the argument above entails that if there are two objects sharing all their NT-properties, they necessarily co-vary with respect to superlative NT-properties like *being the tallest man*. Such properties are unshareable, but if there were objects sharing all their NT-properties, they would necessarily co-vary with respect to them, since they could not possibly have them—in the metaphor of possible worlds, there are no possible worlds where objects sharing all their NT-properties have such superlative NT-properties, and therefore they necessarily co-vary with respect to them. And so, although two objects *a* and *b* that shared all their NT-properties could not share a superlative NT-property, for every NT-property F they would necessarily share the NT-property *necessarily, having F if and only if a has F*. In the next section we shall see, however, that every concrete object must have at least one unshareable NT-property, and therefore objects cannot necessarily co-vary with respect to every NT-property.

Second, someone might object to the argument that descriptive mutual dependence with respect to all properties is absurd, but NMDD does not rule this out, since it only rules out descriptive mutual dependence with respect to all NT-properties. In other words, the objection is that the fact that objects sharing all their NT-properties do not necessarily co-vary with respect to all their properties gives those objects the required descriptive independence, namely descriptive independence with respect to trivializing properties. But this is irrelevant. The argument is cogent in so far as descriptive mutual dependence with respect to all NT-properties is absurd, and I have argued that it is.

Third, nothing like the argument above can support the stronger versions of the Identity of Indiscernibles: PIIa and PIIb. I take it that the corresponding versions of NMDD for pure properties and intrinsic pure properties are true. Indeed, objects that necessarily co-varied with respect to all their pure properties, or even all their intrinsic pure properties, would not have the degree of existential and descriptive ontological autonomy that any concrete object must have. And, of course, it is also necessary that every object x is such that, for every pure property F, whether intrinsic or extrinsic, necessarily, x has F if and only if x has F. Thus, take any concrete object a: it is necessary that a has, for every pure property F, whether intrinsic or extrinsic, the property of *necessarily, having F if and only if a has F*. But this property is not a pure property. Therefore, sharing all the pure properties with object a does not entail having such a property. Thus, it need not be the case that, necessarily, if two concrete objects share all their pure-properties, they necessarily co-vary with respect to every pure property. That is, it need not be the case that the analogue of premise (1) for pure properties holds. Thus, the elements of our argument give us no reason to believe that PIIa, or PIIb for that matter, is true.

Of course, if F is a pure property, whether intrinsic or extrinsic, *necessarily being F* is also a pure property (intrinsic if F is intrinsic and extrinsic if F is extrinsic). Therefore, for every pure property F, objects that share all their pure properties, must also agree with respect to properties like *necessarily being F*, where F is a pure property. That is, if a and b share all their pure properties, it must be the case that either they both have or lack the property of *necessarily being F*, where F is a pure property. But this does not mean that such objects necessarily co-vary with respect to such properties. For instance, a and b could both have the property of *necessarily being spherical*, but they might be such that they need not necessarily co-vary with respect to it, since they might not necessarily coexist. (If you think that nothing could possibly have the property of *necessarily being spherical* unless it existed necessarily, change the example and use the property of *necessarily being spherical if it exists*).

Fourth, it should be clear that the argument from the previous section establishes PII without appealing to any unshareable properties, and so it is an instance of the second type of argument for PII that I distinguished in Section 5.1. Thus, this argument establishes that no two objects can share all their NT-properties without establishing that every object must have a property that it does not share with anything else. Given what this argument shows, it could very well be that, although no two objects shared all their NT-properties, every object shared each one of its NT-properties with some other object.

5.4 I shall now present my second argument for PII. As we shall see, unlike the previous argument, this one appeals to unshareable NT-properties that every concrete object must have.

Take any two objects, *a* and *b*, and suppose they are both green, and therefore they are both coloured. Can they share all their NT-properties? No. For it is necessary that if *a* is green, it has the property of *being coloured because a is green*, while it is necessary that if *b* is green, it has the property of *being coloured because b is green*. But the properties of *being coloured because a is green* and *being coloured because b is green* are distinct properties, and while *a* must have the former property, it cannot have the latter, and while *b* must have the latter property, it cannot have the former. Furthermore, such properties are NT-properties, for differing with respect to them requires differing extra-numerically, since it requires more than merely differing with respect to *being identical with a* or *being identical with b*: it requires differing with respect to the object in virtue of whose greenness one is coloured. Therefore, *a* and *b*, if coloured because green, cannot share all their NT-properties.

This point obviously generalizes, and therefore it is necessary that no two green objects share all their NT-properties. In fact, it generalizes not just to all green objects, or all coloured objects, but to all objects having properties instantiating the determinate/determinable relation. Thus, the fact that not all objects are green, or that not all objects are coloured, shows no limitation in the argument. Uncoloured but massive objects, electrons for instance, will also fail to share all their NT-properties, since no two electrons can share properties like *being massive because electron c has mass m*—such a property can be had only by electron *c*.

I am assuming that having a determinate colour grounds being coloured, and having a determinate mass grounds being massive, and having a determinate shape grounds having a shape, and so on. This seems obvious to me. But I want to emphasize that these assumptions do not mean that determinables can be reduced to determinates, or that all facts about determinables are grounded in determinates, which are much less plausible theses (see Calosi 2021 for an argument against the first thesis, and Wilson 2012 for arguments against both theses). Indeed, I am not even assuming that whenever an object has a determinable it must have at least one of its determinates, an assumption which has been rejected (see Wilson 2013). All I am assuming is that there are certain cases, the majority of cases indeed, in which an object's having a determinable is grounded in that object's having a determinate. This is not a controversial assumption (indeed, some, like Jon Litland (2017: 282, 289) are willing to assume this in full generality, but for my purposes I do not need to make the general assumption). In any case, as we shall now see, my point is even independent of the grounding relations between determinates and determinables.

To see this, note that *existing at time t* is not a determinate of *existing in time*, since existing at time t does not exclude existing at any other time t^*. But such exclusion is necessary between same-level determinates of the same determinables (Johnson 1921: 181).[3] Now, it is necessary that concrete objects exist in time. And, necessarily, any object that exists in time does so in virtue of *its* existing at a particular time. Thus, it is not possible that any concrete objects share all their NT-properties, no matter what kind of objects they are, for if *a* and *b* are two concrete objects, they must differ at least with respect to the following properties: *existing in time because a exists at time t* and *existing in time because*

---

[3] That such exclusion is necessary between same-level determinates of the same determinable has been rejected, for instance by Johansson (2000: 117) and Sanford (2011) (sometimes Armstrong (1978: 113) is interpreted as rejecting such exclusion, e.g. by Wilson (2017), but I think Armstrong can be interpreted as rejecting a different, but related, thesis: that any group of universals exhibiting an intrinsic resemblance order necessarily exclude each other). I am not persuaded by these points against the necessary exclusion between same-level determinates of the same determinable, but this is not the place to discuss the matter.

b exists at time *t*.[4] And it should be clear that such properties are NT-properties, for differing with respect to them requires differing extra-numerically, since it requires more than merely differing with respect to *being identical with a* or *being identical with b*: it requires differing with respect to the object in virtue of whose existence at a particular time one exists in time.

Thus, no matter how strongly causally, spatiotemporally, or qualitatively connected two objects are, it is impossible that one has a determinable because the other one has a corresponding determinate, that one has a general property because the other one has a corresponding specific instance and, in particular, that one exists in time because the other one exists at a particular time. And properties like *being coloured because a is green* and *existing in time because a exists at time t* are NT-properties, since differing with respect to them requires differing extra-numerically.

And thus, since, as I said, it is necessary that concrete objects exist in time, and it is necessary that every object *x* existing in time has the property of *existing in time because x exists at time t*, a property it cannot share with any other object, these considerations generalize to the whole of its intended domain, the domain of concrete objects, and they generalize with necessity. This means that it is impossible that any objects share all their NT-properties, which is what PII claims.

So far, I have presented the argument by means of examples and defended its generalization, but it would be good to formulate it in a more general and abstract way. This is an appropriate formulation:

---

[4] That I think existing at a time *t* does not exclude existing at any other time *t** shows that I am taking existing at a particular time as being weakly located at that time, rather than as being exactly located at it. Following what Parsons says about weak and exact spatial location, we can say that if I am weakly located at time *t*, then *t* is not completely free of me, and if I am exactly located at *t*, then no part of *t* is free of me and every time wholly distinct from *t* is completely free of me (cf. Parsons 2007: 203). Note that (a) to be exactly located at a time *t* excludes being exactly located at any other time *t**, and so it can be taken to be a determinate of existing in time (cf. Calosi 2021) and (b) that objects must exist at some particular time does not mean that they must be exactly located at some particular time; for if every time is a part of another time, then a temporally omnipresent object would be located in time without being exactly located at any time (cf. Parsons 2007: 209).

(1) Necessarily, every object x has the property of *existing in time because x exists at time t*.
(2) Necessarily, no two objects x and y share the property of *existing in time because x exists at time t*.
(3) Necessarily, properties like *existing in time because x exists at time t* are NT-properties.
(4) Therefore, necessarily, no two objects share all their NT-properties (PII).

The argument is valid. Premise (1) is clearly true: (concrete) objects exist in time and they exist in time because of their existing at particular times, and therefore they have properties that express the fact that they exist in time because they exist at a particular time. The truth of premise (2) is what I have been mainly arguing for above. The general point behind the truth of the premise is that when there is a relation of grounding between two properties, an object has the grounded property in virtue of *its* having the grounding property, and not in virtue of some other object having that grounding property. It follows that no objects can share properties like *existing in time because a exists at time t*, which is what premise (2) asserts. I also argued above that properties like *existing in time because x exists at time t* are NT-properties. The reason for this has to do with what differing with respect to them requires. Now, what differing with respect to a property requires is determined by what the property *is*, and since every property is necessarily the property it is, any property such that differing with respect to it requires differing extra-numerically is such that it is necessary that differing with respect to it requires differing extra-numerically. Thus, NT-properties are NT-properties as a matter of necessity, which is what premise (3) asserts of a certain kind of NT-properties.

Thus, there is reason to believe in the conclusion of the argument. In the next section I shall discuss some features of the argument, and some objections to it.

5.5 Let me first note that the argument from the previous section is an instance of the first type of argument for PII that I distinguished in Section 5.1, for it establishes PII by means of unshareable properties.

According to this argument PII is true because, necessarily, every object has an NT-property that it cannot share with any other object.

It is also interesting to note that if every object must have an NT-property that it cannot share with any other object, NMDD must be true. Of course, if one wants to have two independent arguments for PII, one must not base NMDD on the thesis that every object must have an unshareable NT-property—and that is why I provided independent reasons for NMDD.

Let me now discuss a possible objection to the claim that properties like *being coloured because a is green, existing in time because a exists at time t*, and so on, are NT-properties. My reason for this claim was that differing with respect to those properties requires differing extranumerically. This can be supplemented by pointing out that such properties clearly fail to satisfy definition D3 of Section 2.6, according to which trivializing properties must contain at least one property of identity. For the general form of the lambda-expressions of properties like *existing in time because a exists at time t, being coloured because a is green*, and similar ones, is $(\lambda x)(x$ is F because a is G), and there is no open sentence consisting of an identity sign flanked by a constant and a variable in those lambda-expressions. Thus, they contain no properties of identity, and therefore they are NT-properties.

Someone might object that this reveals only that there is a problem with my criterion for trivializing properties. For, the objector will say, a has, but b lacks, the property of *existing in time because a exists at time t* because a is identical with a and b is not, which shows that their difference with respect to the property *existing in time because a exists at time t* consists in their differing with respect to the property *being identical with a*. (The objector could equally use the properties of *being coloured because a is green*, and similar ones, and everything I shall say below will apply to these properties too).

I agree that a and b differ with respect to the property of *existing in time because a exists at time t* because they differ with respect to the property of *being identical with a*. This means, in other words, that their difference with respect to the property *existing in time because a exists at time t* is grounded in their difference with respect to a property of identity. But that a difference with respect to a property F is grounded in a difference with respect to a property of identity does not establish that F is a trivializing property. Indeed, in general, the properties of the ground do not transfer to the grounded. Here are two rather obvious cases: (a) the fact that grounds ground grounded entities does not necessarily make grounded entities grounds; (b) the fact that some non-modal properties or facts ground other properties or facts does not necessarily make the latter non-modal, as those who believe that the modal is grounded in the non-modal will quickly attest. Similarly, that a trivializing property grounds another one does not necessarily make the latter trivializing.

There is another possible objection. Suppose that, necessarily, *a* exists at time *t* if and only if *b* exists at time *t*. Then someone might argue that, if so, *a*'s existence at time *t* is grounded in *b*'s existence at time *t* and *b*'s existence at time *t* is grounded in *a*'s existence at time *t*. Furthermore, the objector will continue, since *a*'s existence in time is grounded in *a*'s existence at time *t*, and *b*'s existence in time is grounded in *b*'s existence at time *t*, it follows by the transitivity of grounding that *a*'s existence in time is grounded in *b*'s existence at time *t*, and *b*'s existence in time is grounded in *a*'s existence at time *t*, and therefore both *a* and *b* have both the properties of *existing in time because a exists at time t* and *existing in time because b exists at time t*. And since this is not peculiar to the properties of *existing in time* and *existing at time t*, it generalizes to all other properties of the form *being F because x is G*, and therefore, the objector will say, it is not the case that every concrete object must have a property of the form *being F because x is G* that it cannot share with any other object.

Now, one weakness of the objection is the reliance on the transitivity of grounding, since this is controversial and has been rejected by some (for instance: Schaffer 2012 and Rodriguez-Pereyra 2015). Another weakness of the objection is the implausible assumption that, necessarily, if *a* exists at time *t* if and only if *b* exists at time *t*, then *a*'s existence at time *t* is grounded in *b*'s existence at time *t* and *b*'s existence at time *t* is grounded in *a*'s existence at time *t*. Such an assumption is initially implausible since, as has been noted so many times in the contemporary discussion on grounding, more is required for grounding than such a merely modal dependence (see, for instance, among many others, Fine 2012: 38 and Sider 2020: 2).

But, even granting those two implausible assumptions, the objection doesn't really damage my argument. For some have drawn a distinction between mediate and immediate grounding (Fine 2012: 50–1), and this is the relevant distinction to apply in this case. Thus, granting the assumptions of the objection, all one should conclude is that *a*'s existence in time is immediately grounded in *a*'s existence at time *t* and mediately grounded in *b*'s existence at time *t*, while *b*'s existence in time is immediately grounded in *b*'s existence at time *t* and mediately grounded in *a*'s existence at time *t*. Indeed, *a* and *b* must differ with respect to the properties of *existing in time immediately because a exists at time t* and *existing in time mediately because b exists at time t*, since *a* must have the former property but *b* must lack both of them. But these properties are no less NT-properties than our original properties of the form *being F because x is G*. Therefore, the objection gives no grounds to doubt that every concrete object must have an NT-property that it cannot share with any other objects.

And this shows the answer to another possible objection. For often an object is coloured because one of its parts is green. More generally, often many of the properties of an object are grounded in the properties of its parts. So, the objection goes, there could be two green objects, *a* and *b*, such that each is a proper part of the other, and each one is coloured because the other one is green. In that case, *a* would have the property of *being coloured because b is green* and *b* would have the property of *being coloured because a is green*. But if the greenness of an object grounds the colouredness of another green object, this is because the greenness of the first object grounds the greenness of the second object. Thus, if green object *a* is coloured because *b* is green, this is because *a* is green because *b* is green, and if green object *b* is coloured because *a* is green, this is because *b* is green because *a* is green. But then *a*'s colouredness will be immediately grounded in its greenness, and mediately grounded in *b*'s greenness, and *b*'s colouredness will be immediately grounded in its greenness, and mediately grounded in *a*'s greenness. Thus, *a* and *b* must differ with respect to the properties of *being coloured immediately because a is green* and *being coloured mediately because b is green*: *a* must have the former property but *b* must lack both of them. And, clearly, the key point generalizes beyond greenness and colouredness: whenever an object's having a determinate or specific property grounds another object's having the corresponding determinable or general property, this is because the first object's having a determinate or specific property grounds the second object's having whatever particular relevant determinate or specific property it has. Therefore, the possibility of mutual parts grounding each other's properties gives no grounds to doubt that every concrete object must have an NT-property that it cannot share with any other objects.

For simplicity I shall from now on ignore the immediately/mediately qualifications on the properties I have been discussing.

5.6 Thus, I have established PII, namely that, necessarily, no two objects share all their NT-properties. And, in fact, one of the arguments by which I have established PII tells us with respect to which properties objects must always differ, namely properties like *existing in time because a exists at time t*, *being coloured because a is green*, and so on. Thus, the spheres in Black's world differ with respect to such properties, and even if they are a counterexample to PIIa and PIIb, they are not a counterexample to PII.

Now, it is easy to see that, necessarily, whenever two objects differ in any way at all, they are weakly discernible in the sense that there is some relation R such that R relates them symmetrically but at least one of them does not bear R to itself. For whenever two objects differ in any way at all, they differ numerically, and numerical difference is a symmetric and irreflexive relation. Thus, the relation of numerical difference weakly discerns any two objects. But the relation of numerical difference corresponds to trivializing properties like *being numerically different from a*, *being numerically different from b*, etc. Thus, a principle stating that, necessarily, every two objects are weakly indiscernible, is subject to trivial proof and therefore it is not a version of the Principle of Identity of Indiscernibles.

But it is also true that whenever two objects differ with respect to *any* NT-property, they are weakly discernible by a relation R that corresponds to NT-properties. This is also easy to see. Consider the relation R such that Rxy if and only if they differ with respect to some NT-property. Such a relation is symmetric and irreflexive, and therefore any two objects that differ with respect to some NT-property are weakly discernible by R. Furthermore, R corresponds to properties like *differing from a with respect to some NT-property, differing from b with respect to some NT-property*, and so on. And such properties are NT-properties.

It can also be seen that if two concrete objects differ with respect to properties like *existing in time because a exists at time t, being coloured because a is green*, etc., they will also be weakly discernible by other relations corresponding to NT-properties. Suppose *a* and *b* differ with respect to the property of *existing in time because a exists at time t*. Consider the relation R such that R*xy* if and only if it is not the case that *x* exists in time because *y* exists at time *t*. And consider two objects *a* and *b* that exist at time *t*. Since neither *a* exists in time because *b* exists at time *t*, nor does *b* exist in time because *a* exists at time *t*, *a* and *b* are symmetrically related by R. Furthermore, neither object bears R to itself, since *a* exists in time because *a* exists at time *t*, and *b* exists in time because *b* exists at time *t*. Therefore, R weakly discerns them. And R corresponds to NT-properties like *being such that it does not exist in time because a exists at time t, being such that it does not exist in time because b exists at time t*, etc.[5]

The general point is that, as we saw above, whenever two objects differ with respect to any NT-property, they are weakly discernible by a relation that corresponds to NT-properties. And since I have made no contingent assumptions in my reasoning, this holds of necessity. This suggests the following alternative formulation of PII, where being *non-trivially* weakly discernible means being weakly discernible by a relation corresponding to NT-properties:

PII: Necessarily, every two objects are non-trivially weakly discernible.

But if this is an alternative formulation of what I have argued for, how does my project relate to the interesting and influential project of

---

[5] Note that, even if both *a* and *b* are concrete, in the case of properties like *being coloured because a is green* it might be that only one of them fails to be related by the relation R such that R*xy* if and only if it is not the case that *x* is coloured because *y* is green. For suppose that *a* and *b* differ with respect to the property of *being coloured because a is green*, but *b* is either not coloured or not green. Then *a* fails to bear R to itself, but *b* bears it to itself, since it is not the case that *b* is coloured because *b* is green. Never mind: *a* and *b* will still be weakly discernible by R.

philosophers like Saunders (2003, 2006) and Muller (2015) who have argued that all that is needed for discernibility is weak discernibility and that Black's spheres and physical particles are weakly discernible and therefore satisfy PII?

There are two important differences between what I have done and what Saunders, Muller, and others do. The first one is that I have explained what is required for a version of PII not to be subject to trivial proof and what trivializing and non-trivializing properties are. True, Saunders and Muller don't defend a trivial version of PII, since the weakly discernible relations they use to discern allegedly indiscernible objects correspond to NT-properties. But they do not propose a definition or even a criterion for NT-properties or relations and, in fact, their remarks suggest that they take any impure property to be trivializing (Muller 2015: 227).

The second difference is that Saunders and Muller do not offer arguments for PII. What they offer is a defence of PII from alleged counterexamples—they defend it by showing that the alleged indiscernible objects are in fact weakly discernible (Saunders 2003, 2006, Saunders and Muller 2008, Muller 2015). Instead, I have given two arguments that PII must hold true.

Let me expand on this second difference. First, note that since Saunders and Muller appeal to *physical* relations to counter the alleged counterexamples to PII, they are not in a position to defend PII from all possible counterexamples since some of the possible counterexamples do not involve physical objects. Instead, having established PII, I have thereby established that there is no possible counterexample to it and, by means of my second argument, I am in a position to know with respect to which kinds of properties any two objects must differ. Let us see this in more detail.

Saunders and Muller point out that Black's spheres are weakly discernible by a physical relation like *being two miles away from* (Saunders 2006: 57, Muller 2015: 210). But consider Della Rocca's point that, if Black's spheres are indiscernible, there is no reason to deny that there could be 20 co-located spheres, that is, 20 spheres occupying exactly the same place at the same time (Della Rocca 2005: 486). Muller's response is that Della Rocca has overlooked the possibility of discerning Black's spheres relationally (Muller 2015: 218). But what if there could be 20 co-located spheres, whether or not Black's spheres are indiscernible? Although Saunders does not consider this case, he considers what is basically the same case, the case of co-located classical particles. And he says that co-located particles with zero-velocity would not be weakly discernible and therefore, unless a more refined description is available, we should say that there is only one single particle there (Saunders 2006: 60). But there is always a more refined description available: the co-located particles, and the co-located spheres, will differ with respect to properties like *existing in time because a exists at time t*, and similar ones.

Indeed, it is interesting to note that the relation *being two miles away from* is not necessarily irreflexive, since in spaces with sufficient curvature objects can bear that relation to themselves. And there could be a situation in which there are two spheres that are co-located at two places two miles from each other. In that case the spheres are symmetrically and reflexively related by the relation *being two miles away from*, and therefore they are not weakly discernible through that relation. Nevertheless, even in that situation the spheres will differ with respect to properties like *existing in time because a exists at time t, being coloured because a is green*, and so on.

Furthermore, Saunders and Muller think that elementary bosons are not weakly discernible (Saunders 2006: 60, Saunders and Muller 2008: 542). They do not see this as a counterexample to PII because they think that, by failing to satisfy PII, bosons are not objects. But this is due to a terminological decision about the word 'object' (Saunders and Muller 2008: 503; cf. Caulton and Butterfield 2012: 31). Without that terminological decision, they could equally have said that elementary bosons are indiscernible objects. But Muller and Seevinck have then gone on to argue that it follows from certain quantum mechanical postulates that all particles, including elementary bosons, are weakly discernible (Muller and Seevinck 2009: 181). If they are right, no physical particles violate PII. But this does not amount to having established PII, for it does not establish that, necessarily, no two objects share all their NT-properties. Indeed, it does not even establish the truth of PII for the actual world since, for all Muller and Seevinck have shown, physical particles that refuted PII could yet be discovered. All they have done is to show that there are no known physical particles that are a counterexample to PII. That is no small deed. But we can go beyond that. No physical particles *a* and *b*, whether fermions, bosons, or what have you, could share all their NT-properties, since they must differ with respect to the property of *existing in time because a exists at time t*. And some if not all of them must also differ with respect to other properties too. For instance, if *a* and *b* are photons, they must also differ with respect to the property of *having angular momentum because b has spin 1*.

There are many other cases which are *prima facie* counterexamples to PII, and for which the vagaries of physical contingency are of no help. Consider two immaterial *disembodied* minds, *a* and *b*, which have exactly the same pure properties and bear no mental relation to each other—that is, they have never thought nor will ever think about each other. And suppose they are both thinking the thought that 6 is a perfect number. One might have thought that such minds share all their NT-properties. But they don't: they differ, for instance, with respect to the property of *thinking about a number because b is thinking that 6 is a perfect number*.

5.7 As I said in Section 1.2, I decided to restrict PII to concrete objects. I also said there that one of my arguments for PII for concrete objects could also be used to prove a more general version of PII, one that says that, necessarily, no two objects, whether abstract or concrete, can share all their NT-properties. This is because abstract objects, like concrete ones, must have complex properties of the form *being F because x is G*. And every two abstract objects will differ with respect to some such properties.

Indeed, all abstract objects, no matter what kind of abstract objects they are, must have some property of the form *being F because x is G*, and every two abstract objects must differ with respect to some such properties. For every abstract object, of whatever kind, must have the property of *being abstract or concrete*, and it must have that property because *it* is abstract. Thus, every two abstract objects *a* and *b* must therefore differ with respect to the property of *being abstract or concrete because a is abstract*: *a* must have such a property, and *b* must lack it. Thus, every two abstract objects, of whatever kind, must differ in at least one property (from which it follows, furthermore, that all abstract objects, of whatever kind, must satisfy NMDD). Therefore PII can be extended to apply to abstract objects too.

We can see this in an example that has been thought to show that some abstract objects fail to satisfy PII. According to Leitgeb and Ladyman, graph nodes can fail to satisfy PII. But such nodes obey PII. Consider any two such nodes, a and b. No matter what graph they appear in, they will have different properties: one will have the property of *being a graph-theoretic object because a is a node* while the other will have the different property of *being a graph-theoretic object because b is a node*.

So far so good, but the fact is that Leitgeb and Ladyman attempted to counterexemplify PII by means of *unlabelled* nodes. Indeed, they argue in particular that the unlabelled nodes of a two-node graph without an edge fail to be even weakly discernible and are completely indiscernible despite being *two* nodes (Leitgeb and Ladyman 2008: 392). Now, Rafael De Clercq has argued that nothing here is a counterexample to PII because so-called unlabelled graphs are not really graphs but rather isomorphism classes of labelled graphs (de Clercq 2012: 671). As far as I can tell, De Clercq may well be right about this (cf. the discussion in Duguid 2016). But I am not going to enter into that discussion, since I think one can respond to Leitgeb and Ladyman even if one concedes to them that unlabelled graphs are really graphs.

In effect, the appeal to *unlabelled* nodes makes no difference at all. We can assign one of the nodes to a variable and the other node to another variable. Indeed, the nodes being unlabelled, we can only resort to variables to describe the graph, and this is in fact how Leitgeb and Ladyman describe it: there is an x and there is a y, such that both x and y are nodes, and x and y are numerically different and every node is either identical to x or identical to y (Leitgeb and Ladyman 2008: 394). There are no free variables in this description, but for this description to be true the variables 'x' and 'y' must be assigned different nodes. But then, once such an assignment has provided the variables' denotation, 'is a graph-theoretic object because x is a node' and 'is a graph-theoretic object because y is a node' express different properties (remember that impure properties can be expressed by means of variables: see Section 1.8).

But then it is clear that the node assigned to 'x' has the property of *being a graph-theoretic object because x is a node*, while the node assigned to 'y' has the property of *being a graph-theoretic object because y is a node*, and they do not share these properties. Thus, the two nodes satisfy PII—indeed they are non-trivially weakly indiscernible in the sense defined in the previous section.

This is just an example, although a particularly illustrative one. The main point is the one I made above: PII holds true not only of concrete objects but of abstract objects too, and therefore it holds true of all objects whatsoever.

References

Adams, R. M. 1979. 'Primitive Thisness and Primitive Identity', *The Journal of Philosophy* 6, pp. 5–26.
Alvarado, J. T. 2020. *A Metaphysics of Platonic Universals and their Instantiations*, Cham, Switzerland: Springer.
Armstrong, D. M. 1978. *A Theory of Universals. Universals and Scientific Realism*, Volume II, Cambridge: Cambridge University Press.
Assadian, B. 2019. 'In Defense of Utterly Indiscernible Entities', *Philosophical Studies* 176, pp. 2551–61.
Audi, P. 2011. 'Primitive Causal Relations and the Pairing Problem', *Ratio* 24, pp. 1–16.
Ayer, A. J. 1954. 'The Identity of Indiscernibles', in A. J. Ayer, *Philosophical Essays*, London: Macmillan.
Bader, R. 2013. 'Towards a Hyperintensional Theory of Intrinsicality', *The Journal of Philosophy* 110, pp. 525–63.
Baldwin, T. 1996. 'There Might Be Nothing', *Analysis* 56, pp. 231–8.
Bennett, K. 2004. 'Spatio-temporal Coincidence and the Grounding Problem', *Philosophical Studies* 118, pp. 339–71.
Black, M. 1952. 'The Identity of Indiscernibles', *Mind* 61, pp. 153–64.
Bohn, E. D. 2014. 'From Hume's Dictum via Submergence to Composition as Identity or Mereological Nihilism', *Pacific Philosophical Quarterly* 95, pp. 336–55.
Brody, B. 1980. *Identity and Essence*, Princeton: Princeton University Press.
Burgess, A. 2012. 'A Puzzle about Identity', *Thought* 1, pp. 90–9.
Calosi, C. 2021. 'Determinables, Location, and Indeterminacy', *Synthese* 198, pp. 4191–4204.
Calosi, C. and Varzi, A. 2016. 'Back to Black', *Ratio* 29, pp. 1–10.
Carnap, R. 1988. *Meaning and Necessity*, Chicago: The University of Chicago Press.
Caulton, A. and Butterfield, J. 2012. 'On Kinds of Indiscernibility in Logic and Metaphysics', *The British Journal for the Philosophy of Science* 63, pp. 27–84.
Cortes, A. 1976. 'Leibniz's Principle of the Identity of Indiscernibles: A False Principle', *Philosophy of Science* 43, pp. 491–505.
Cowling, S. 2015. 'Non-qualitative Properties', *Erkenntnis* 80, pp. 275–301.
Cowling, S. 2017. *Abstract Entities*, Abingdon and New York: Routledge.
Cross, C. B. 2011. 'Brute Facts, the Necessity of Identity, and the Identity of Indiscernibles', *Pacific Philosophical Quarterly* 92, pp. 1–10.
Curtis, B. 2014. 'The Rumble in the Bundle', *Noûs* 48, pp. 298–313.
De Clercq, R. 2012. 'On some Putative Graph-theoretic Counterexamples to the Principle of the Identity of Indiscernibles', *Synthese* 187, pp. 661–76.

Della Rocca, M. 2005. 'Two Spheres, Twenty Spheres, and the Identity of Indiscernibles', *Pacific Philosophical Quarterly* 86, pp. 480–92.
Dorato, M. and M. Morganti. 2013. 'Grades of Individuality. A Pluralistic View of Identity in Quantum Mechanics and in the Sciences', *Philosophical Studies* 163, pp. 591–610.
Duguid, C. J. 2016. 'Graph Theory and the Identity of Indiscernibles', *Dialectica*, 70, pp. 463–74.
Eddon, M. 2011. 'Intrinsicality and Hyperintensionality', *Philosophy and Phenomenological Research* 82, pp. 314–36.
Erdrich, N. 2020. 'Répliques, doubles, et coïncidents purs. Trois régimes d'indiscernables', *Philosophia Scientiae* 24, pp. 29–52.
Ereshefsky, M. 2017. 'Species', *The Stanford Encyclopedia of Philosophy* (Fall 2017 Edition), Edward N. Zalta (ed.), URL = <https://plato.stanford.edu/archives/fall2017/entries/species/>.
Figdor, C. 2008. 'Intrinsically/Extrinsically', *The Journal of Philosophy* 105, pp. 691–718.
Fine, K. 2000. 'A Counter-Example to Locke's Thesis', *The Monist* 83, pp. 357–61.
Fine, K. 2012. 'Guide to Ground', in F. Correia and B. Schnieder (eds), *Metaphysical Grounding: Understanding the Structure of Reality*, Cambridge: Cambridge University Press, 2012, pp. 37–80.
Francescotti, R. 1999. 'How to Define Intrinsic Properties', *Noûs* 33, pp. 590–609.
French, S. 1989. 'Why the Principle of the Identity of Indiscernibles is not Contingently True Either', *Synthese* 78, pp. 141–66.
French, S. 1995. 'Hacking away at the Identity of Indiscernibles: Possible Worlds and Einstein's Principle of Equivalence', *The Journal of Philosophy* 92, pp. 455–66.
Goodman, J. 2015. 'Consequences of conditional excluded middle', unpublished, https://jeremy-goodman.com/ retrieved on 7 September 2020.
Hacking, I. 1975. 'The Identity of Indiscernibles', *The Journal of Philosophy* 72, pp. 249–56.
Hale, B. 1996. 'Absolute Necessities', *Philosophical Perspectives* 10, pp. 93–117.
Hale, B. 2015. *Necessary Beings*, Oxford: Oxford University Press.
Hawley, K. 2009. 'Identity and Indiscernibility', *Mind* 469, pp. 101–19.
Hazen, A. 1979. 'Counterpart-Theoretic Semantics for Modal Logic', *The Journal of Philosophy* 76, pp. 319–38.
Heathcote, A. 2022. 'Five Indistinguishable Spheres', *Axiomathes* 32, pp. 367–83.
Hoy, R. C. 1984. 'Inquiry, Intrinsic Properties, and the Identity of Indiscernibles', *Synthese* 61, pp. 275–97.
Hughes, C. 1999. 'Bundle Theory from A to B', *Mind* 108, pp. 149–56.
Johansson, I. 2000. 'Determinables as Universals', *The Monist* 83, pp. 101–21.
Johnson, W. E. 1921. *Logic*. Part 1, Cambridge: Cambridge University Press.
Jones, N. K. 2016. 'A Higher-Order Solution to the Problem of the Concept *Horse*', *Ergo: An Open Access Journal of Philosophy* 3, pp. 132–66.
Jubien, M. 2009. *Possibility*. Oxford: Oxford University Press.
Katz, B. 1983. 'The Identity of Indiscernibles Revisited', *Philosophical Studies* 44, pp. 37–44.

Khamara, E. J. 1988. 'Indiscernibles and the Absolute Theory of Space and Time', *Studia Leibnitiana*, 20, pp. 140–59.
King, P. 2000. 'The Problem Of individuation in the Middle Ages', *Theoria* 66, pp. 159–84.
Kripke, S. 1981. *Naming and Necessity*, Oxford: Blackwell Publishers Ltd.
Koslicki, K. 2018. *Form, Matter, Substance*, Oxford: Oxford University Press.
Ladyman, J. 2005. 'Mathematical Structuralism and the Identity of Indiscernibles', *Analysis* 65, pp. 218–21.
Ladyman, J. and T. Bigaj. 2010. 'The Principle of Identity of Indiscernibles and Quantum Mechanics', *Philosophy of Science*, 2010, pp. 117–36.
Ladyman, J., Ø. Linnebo, and R. Pettigrew. 2012. 'Identity and Discernibility in Philosophy and Logic', *The Review of Symbolic Logic* 5, pp. 162–86.
Langton, R. and D. Lewis. 1998. 'Defining "Intrinsic"', *Philosophy and Phenomenological Research* 58, pp. 333–45.
Langton, R. and D. Lewis. 2001. 'Marshall and Parsons on "Intrinsic"', *Philosophy and Phenomenological Research* 63, pp. 353–5.
Legenhausen, G. 1989. 'Moderate Anti-Haecceitism', *Philosophy and Phenomenological Research* 49, pp. 625–42.
Leibniz, G. W. 1956. *The Leibniz–Clarke Correspondence*, edited by H. G. Alexander, Manchester: Manchester University Press.
Leibniz, G. W. 1967. *The Leibniz–Arnauld Correspondence*, edited and translated by H. T. Mason, Manchester: Manchester University Press.
Leibniz, G. W. 2020. *Discourse on Metaphysics*, translated with introduction and commentary by G. Rodriguez-Pereyra, Oxford: Oxford University Press.
Leitgeb, H. and J. Ladyman. 2008. 'Criteria of Identity and Structuralist Ontology', *Philosophia Mathematica* 16, pp. 388–96.
Lewis, D. 1973, *Counterfactuals*. Oxford: Blackwell.
Lewis, D. 1983a. 'Extrinsic Properties', *Philosophical Studies* 44, pp. 197–200.
Lewis, D. 1983b. 'New Work for a Theory of Universals', *Australasian Journal of Philosophy* 61, pp. 343–77.
Lewis, D. 1983c. 'Postscripts to "Counterpart Theory and Quantified Modal Logic"', in D. Lewis, *Philosophical Papers*, vol. 1., New York and Oxford: Oxford University Press, 1983.
Lewis, D. 1986. *On the Plurality of Worlds*, Oxford and Cambridge, MA: Blackwell.
Lewis, D. 1991. *Parts of Classes*, Oxford: Basil Blackwell.
Litland, J. 2017. 'Grounding Ground', *Oxford Studies in Metaphysics* 10, pp. 279–315.
MacBride, F. 2006. 'What Constitutes the Numerical Diversity of Mathematical Objects?', *Analysis* 66, pp. 63–9.
McMichael, A. 1983. 'A New Actualist Modal Semantics', *Journal of Philosophical Logic* 12, pp. 73–99.
McTaggart, J. M. E. 1921. *The Nature of Existence*, volume 1, Cambridge: Cambridge University Press.
Markosian, N. 1998. 'Brutal Composition', *Philosophical Studies* 92, pp. 211–49.
Marshall, D. and J. Parsons. 2001. 'Langton and Lewis on "Intrinsic"', *Philosophy and Phenomenological Research* 63, pp. 347–52.

Marshall, D. and B. Weatherson. 2018. 'Intrinsic vs. Extrinsic Properties', *The Stanford Encyclopedia of Philosophy* (Spring 2018 Edition), Edward N. Zalta (ed.), URL=<https://plato.stanford.edu/archives/spr2018/entries/intrinsic-extrinsic/>.

Muller, F. A. 2015. 'The Rise of Relationals', *Mind* 124, pp. 201–37.

Muller, F. A. and M. P. Seevinck. 2009. 'Discerming Elementary Particles', *Philosophy of Science* 76, 179–200.

O'Connor, D. J. 1954. 'The Identity of Indiscernibles', *Analysis* 14, pp. 103–10.

O'Leary-Hawthorne, J. 1995. 'The Bundle Theory of Substance and the Identity of Indiscernibles', *Analysis* 55, pp. 191–6.

Odegard, D. 1964. 'Indiscernibles', *The Philosophical Quarterly* 14, pp. 204–13.

Parsons, J. 2007. 'Theories of Location', *Oxford Studies in Metaphysics* 3, pp. 201–32.

Plate, J. 2022. 'Qualitative Properties and Relations', *Philosophical Studies* 179, pp. 1297–1322.

Quine, W. V. O. 1960. *Word and Object*, Cambridge, MA: Harvard University Press.

Quine, W. V. O. 1976. 'Grades of Discriminability', *The Journal of Philosophy* 73, pp. 113–16.

Quine, W. V. O. 2000. 'On what there is', in J. Kim and E. Sosa (eds), *Metaphysics. An Anthology*, Malden, MA and Oxford: Blackwell, 2000, pp. 4–12.

Robinson, D. 2000. 'Identities, Distinctnesses, Truthmakers, and Indiscernibility Principles', *Logique et Analyse* 43, pp. 145–83.

Rodriguez-Pereyra, G. 1997. 'There Might Be Nothing: the Subtraction Argument Improved', *Analysis* 57, pp. 159–66.

Rodriguez-Pereyra, G. 2002. *Resemblance Nominalism*. Oxford: Oxford University Press.

Rodriguez-Pereyra, G. 2004. 'The Bundle Theory is Compatible with Distinct but Indiscernible Particulars', *Analysis* 64, pp. 72–81.

Rodriguez-Pereyra, G. 2006. 'How not to trivialize the Identity of Indiscernibles', in P. F. Strawson and A. Chakrabati (eds), *Universals, Concepts and Qualities. New Essays on the Meaning of Predicates*. Aldershot, Hants, and Burlington, VT: Ashgate, pp. 205–23.

Rodriguez-Pereyra, G. 2014. *Leibniz's Principle of Identity of Indiscernibles*. Oxford: Oxford University Press.

Rodriguez-Pereyra, G. 2015. 'Grounding is not a strict order', *Journal of the American Philosophical Association* 1, pp. 517–34.

Rodriguez-Pereyra, G. 2017a. 'The Argument from almost Indiscernibles', *Philosophical Studies* 174, pp. 3005–20.

Rodriguez-Pereyra, G. 2017b. 'Indiscernible Universals', *Inquiry* 60, pp. 604–24.

Rodriguez-Pereyra, G. 2018. 'The Principles of Contradiction, Sufficient Reason, and Identity of Indiscernibles', in M. R. Antognazza (ed.), *The Oxford Handbook of Leibniz*, New York: Oxford University Press, pp. 45–64.

Sanford, D. 2011. 'Determinables and Determinates', *The Stanford Encyclopedia of Philosophy* (Spring 2011 Edition), Edward N. Zalta (ed.), <https://plato.stanford.edu/archives/spr2011/entries/determinate-determinables/>.

Saunders, S. 2003. 'Physics and Leibniz's Principles', in K. Brading and E. Castellani (eds), *Symmetries in Physics*, Cambridge: Cambridge University Press, pp. 289–307.

Saunders, S. 2006. 'Are Quantum Particles Objects?', *Analysis* 66, pp. 52–63.

Saunders, S. and F. Muller. 2008. 'Discerning Fermions', *The British Journal for the Philosophy of Science* 59, pp. 499–548.
Schaffer, J. 2004. 'Two Conceptions of Sparse Properties', *Pacific Philosophical Quarterly* 85, pp. 92–102.
Schaffer, J. 2012. 'Grounding, Transitivity, and Contrastivity', in F. Correia and B. Schnieder (eds), *Metaphysical Grounding: Understanding the Structure of Reality*, Cambridge: Cambridge University Press, pp. 122–38.
Shiver, A. 2014. 'Mereological Bundle Theory and the Identity of Indiscernibles', *Synthese* 191, pp. 901–13.
Shumener, E. 2017. 'The Metaphysics of Identity: is Identity Fundamental?', *Philosophy Compass* 12: e12397, doi: 10.1111/phc3.12397, pp. 1–13.
Shumener, E. 2021. 'Do Identity and Distinctness Facts Threaten the PSR?', *Philosophical Studies* 178, pp. 1023–41.
Sider, T. 1996. 'Intrinsic Properties', *Philosophical Studies* 83, pp. 1–27.
Sider, T. 2001. 'Maximality and Intrinsic Properties', *Philosophy and Phenomenological Research* 63, pp. 357–64.
Sider, T. 2006. 'Bare Particulars', *Philosophical Perspectives* 20, pp. 387–97.
Sider, T. 2020. *The Tools of Metaphysics and the Metaphysics of Science*, Oxford: Oxford University Press.
Trueman, R. 2021. *Properties and Propositions. The Metaphysics of Higher Order Logic*, Cambridge: Cambridge University Press.
Valicella, W. 1997. 'Bundles and Indiscernibility: A Reply to O'Leary-Hawthorne', *Analysis* 57, pp. 91–4.
van Cleve, J. 1985. 'Three Versions of the Bundle Theory', *Philosophical Studies* 47, pp. 95–107.
van Cleve, J. 2002. 'Time, Idealism and the Identity of Indiscernibles', *Philosophical Perspectives* 16, pp. 379–93.
van Inwagen, P. 2002. 'The Number of Things', *Philosophical Issues* 12, pp. 176–96.
van Inwagen, P. 2004. 'A Theory of Properties', in D. Zimmerman (ed.), *Oxford Studies in Metaphysics*, vol. 1, Oxford: Oxford University Press, pp. 107–38.
Wang, J. 2015. 'Actualist Counterpart Theory', *The Journal of Philosophy* 112, pp. 417–41.
Weatherson, B. 2001. 'Intrinsic Properties and Combinatorial Principles', *Philosophy and Phenomenological Research* 63, pp. 365–80.
Whitehead, A. N. and B. Russell 1925. *Principia Mathematica*, volume 1, Cambridge: Cambridge University Press.
Wiggins, D. 2016. *Continuants. Their Activity, Their Being and Their Identity*, Oxford: Oxford University Press.
Wilson, J. 2010. 'Hume's Dictum, and Why Believe It?', *Philosophy and Phenomenological Research* 80, pp. 595–637.
Wilson, J. 2012. 'Fundamental Determinables', *Philosophers' Imprint* 12, pp. 1–17.
Wilson, J. 2013. 'A Determinable-Based Account of Metaphysical Indeterminacy', *Inquiry* 56, pp. 359–85.
Wilson, J. 2017. 'Determinables and Determinates', *The Stanford Encyclopedia of Philosophy* (Spring 2017 Edition), Edward N. Zalta (ed.), <https://plato.stanford.edu/archives/spr2017/entries/determinate-determinables/>.

Wörner, D. 2021. 'On making a difference: towards a minimally non-trivial version of the Identity of Indiscernibles', *Philosophical Studies* 12, pp. 4261–78.

Wright, C. 2001. 'Why Frege does not deserve his grain of salt', in B. Hale and C. Wright, *The Reason's Proper Study. Essays Towards a Neo-Fregean Philosophy of Mathematics*, Oxford: Oxford University Press, 2001, pp. 72–90.

Yablo, S. 1987. 'Identity, Essence, and Indiscernibility', *The Journal of Philosophy* 84, pp. 293–314.

Yablo, S. 1990. 'The Real Distinction between Mind and Body', *Canadian Journal of Philosophy* Supp. Vol. 16, pp. 149–201.

Zimmerman, D. 1997. 'Distinct Indiscernibles and the Bundle Theory', *Mind* 106, pp. 305–9.

Index

For the benefit of digital users, indexed terms that span two pages (e.g., 52–53) may, on occasion, appear on only one of those pages.

Actualism 17–18, 69, 81–2, 95–6
Adams, R. M. 21–2, 54–5, 76–7

Black, M. 61–2, 69n.4, 72n.6, 73n.7, 75
    see also Black's world
Black's world 61, 75

Calosi 77

Della Rocca, M. 54–5, 121–2
Determinates and determinables 112–14, 118–19
Difference, extra-numerical 5–7, 12–13, 19, 27n.9, 31, 109–10, 112–15

Goodman, J. 69–72
Ground 112–19, 121–5

Hacking, I. *see* hacking's objection
Hacking's objection 75–7, 81, 83, 86, 95–8
Haecceities 5–6
Hoy, R. 68–70
Hume's dictum 82, 101–5

Identity, primitive 54–5
Indiscernibility of Identicals 7
Individual essences 23, 41–3, 101
Individuation 56–9, 84

Katz, B. 31, 37–8, 103
Kripke, S. 90–1

Ladyman, J. 124–5
Lambda-operator 16, 40
Leibniz, G. W. 5n.1, 56, 66n.2, 67, 101

Leitgeb, H. 124–5
Lewis, D. 11, 47–8, 79–82

Modal realism 10n.3, 18n.5, 79–82, 89n.8, 95–6
Muller, F. 72–4, 120–3

Necessity, metaphysical 8

Objects 9
    abstract 13, 66, 90, 123–5
    concrete 13–14

Possible worlds, *see* Actualism and Modal Realism
Properties 9–13
    modal 17–18, 68–72
    intrinsic and extrinsic 27–30, 84–7, 111
    of identity and difference 5–7, 12, 27–8, 116–17
    pure and impure 21–7, 45, 84–6, 92, 94, 111
    relational 19–21
    superlative 45, 110
    temporal 16–17
    trivializing and non-trivializing 6, 31, 101

Quine, W. V. O. 20, 26–7

Relations 18–19, 72–4

Saunders, S. 120–3
Shiver, A. 98
Substance 8, 103–4

Subtraction argument, *see* Subtraction Principle
Subtraction Principle 88, 90–2, 98–100

Universal quantifier, first-order and second-order 8–10, 13–15

Varzi, A. 77

Weak Discernibility 20, 119–22
Wilson, J. 102, 113, 113n.3

Yablo, S. 106n.2